Honda CBX550
Owners
Workshop
Manual

by Pete Shoemark

Models covered
CBX550 FC and F2C. 572.5cc. February 1982 to April 1985
CBX550 FD and F2D. 572.5cc. January 1984 to September 1986

ISBN 978 0 85696 940 9

Printed in the UK *(940-1AQ8)*

A B C D E
F G H I J
K L M N O
P Q R S
2

Haynes Publishing
Sparkford Nr Yeovil
Somerset BA22 7JJ England

Haynes Publishing
859 Lawrence Drive
Newbury Park
California 91320 USA

Acknowledgements

Our thanks are due to Paul Branson Motorcycles of Yeovil, who supplied the Honda CBX550 featured in the photographs throughout this manual. The Avon Rubber Company supplied information and technical assistance on tyre care and fitting, and NGK Spark Plugs (UK) Ltd provided information on plug maintenance and electrode conditions.

About this manual

The purpose of this manual is to present the owner with a concise and graphic guide which will enable him to tackle any operation from basic routine maintenance to a major overhaul. It has been assumed that any work would be undertaken without the luxury of a well-equipped workshop and a range of manufacturer's service tools.

To this end, the machine featured in the manual was stripped and rebuilt in our own workshop, by a team comprising a mechanic, a photographer and the author. The resulting photographic sequence depicts events as they took place, the hands shown being those of the author and the mechanic.

The use of specialised, and expensive, service tools was avoided unless their use was considered to be essential due to risk of breakage or injury. There is usually some way of improvising a method of removing a stubborn component, providing that a suitable degree of care is exercised.

The author learnt his motorcycle mechanics over a number of years, faced with the same difficulties and using similar facilities to those encountered by most owners. It is hoped that this practical experience can be passed on through the pages of this manual.

Where possible, a well-used example of the machine is chosen for the workshop project, as this highlights any areas which might be particularly prone to giving rise to problems. In this way, any such difficulties are encountered and resolved before the text is written, and the techniques used to deal with them can be incorporated in the relevant section. Armed with a working knowledge of the machine, the author undertakes a considerable amount of research in order that the maximum amount of data can be included in the manual.

A comprehensive section, preceding the main part of the manual, describes procedures for carrying out the routine maintenance of the machine at intervals of time and mileage. This section is included particularly for those owners who wish to ensure the efficient day-to-day running of their motorcycle, but who choose not to undertake overhaul or renovation work.

Each Chapter is divided into numbered sections. Within these sections are numbered paragraphs. Cross reference throughout the manual is quite straightforward and logical. When reference is made 'See Section 6.10' it means Section 6, paragraph 10 in the same Chapter. If another Chapter were intended, the reference would read, for example, 'See Chapter 2, Section 6.10'. All the photographs are captioned with a section/paragraph number to which they refer and are relevant to the Chapter text adjacent.

Figures (usually line illustrations) appear in a logical but numerical order, within a given Chapter. Fig. 1.1 therefore refers to the first figure in Chapter 1.

Left-hand and right-hand descriptions of the machines and their components refer to the left and right of a given machine when the rider is seated normally.

Motorcycle manufacturers continually make changes to specifications and recommendations, and these, when notified, are incorporated into our manuals at the earliest opportunity.

We take great pride in the accuracy of information given in this manual, but motorcycle manufacturers make alterations and design changes during the production run of a particular motorcycle of which they do not inform us. No liability can be accepted by the authors or publishers for loss, damage or injury caused by any errors in, or omissions from, the information given.

Contents

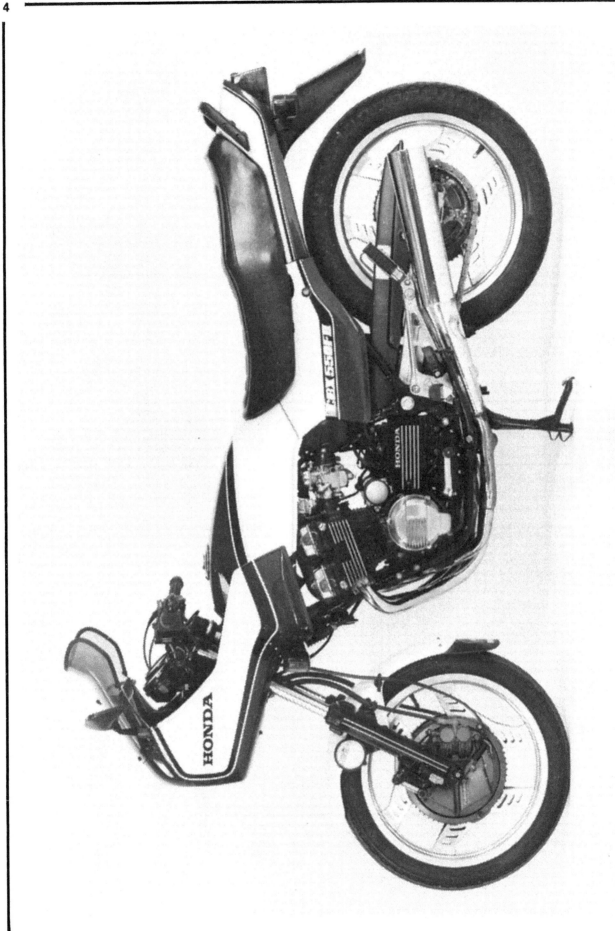

Left-hand view of the Honda CBX550 F2

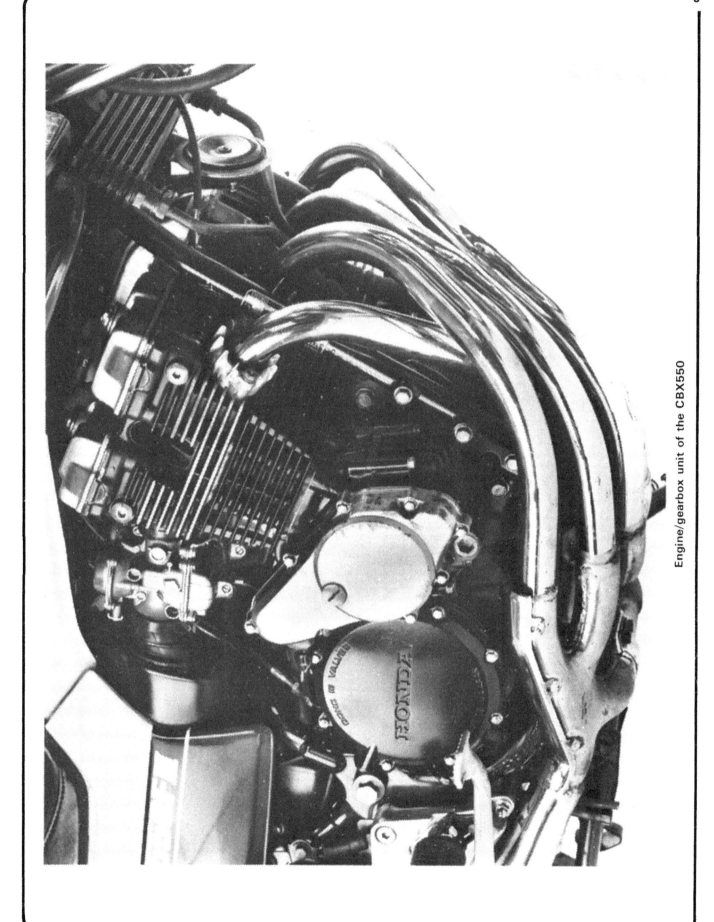

Engine/gearbox unit of the CBX550

Introduction to the Honda CBX550

The CBX550 is representative of a new generation of four-cylinder motorcycles, combining lightness and agility and remarkable performance in a relatively small machine. Powered by a conventional in-line four cylinder engine, the machine might appear rather uninspiring alongside some of the more exotic V-engined offerings, but whilst the engine unit is of familiar layout, its characteristics are closer to those of recent 750 cc machines, offering a high top speed but retaining a smooth spread of power even at fairly low engine speeds.

The chassis incorporates a number of recent technical innovations, the most striking being the inboard ventilated disc brakes. A drum enclosure protects the ventilated discs from rain and permits the use of cast iron as the disc material without the attendant cosmetic problems of rusting. The discs are unusual in motorcycle applications, the outer edge forming the mounting points whilst the caliper grips from the inner edge. The caliper units are of twin piston design, both pistons acting from one side, and are mounted on a ventilated backplate which shrouds the discs themselves.

Front suspension is by oil-damped telescopic fork with linked air assistance. The left-hand fork leg incorporates an anti-dive mechanism which reacts to torque loadings to increase damping resistance under heavy braking. The machine is thus held more stably under these conditions than would otherwise be the case. To encourage equal movement of both forks, a forged alloy fork brace joins and stiffens the two lower legs.

Rear suspension is by Pro-link, a Honda designed rising rate system which allows compliant, long travel suspension which is also capable of maintaining control of the wheel over more arduous road surfaces, the springing and damping forces increasing in proportion to rear wheel movement. The single central suspension unit also features air assistance to allow fine-tuning of the suspension to suit the rider's preference.

Two machines are available, the CBX550 F, and the CBX550 F2 which is identical but for the addition of a half-fairing.

Model dimensions and weights

	CBX550 F	CBX550 F2
Overall length	2085 mm (82.1 in)	2085 mm (82.1 in)
Overall width	740 mm (29.1 in)	740 mm (29.1 in)
Overall height	1080 mm (42.5 in)	1270 mm (50.0 in)
Wheelbase	1380 mm (54.3 in)	1380 mm (54.3 in)
Seat height	785 mm (30.9 in)	785 mm (30.9 in)
Ground clearance	140 mm (5.5 in)	140 mm (5.5 in)
Dry weight	184 kg (406 lb)	190 kg (419 lb)

Ordering Spare Parts

When ordering spare parts for any Honda model it is advisable to deal direct with an official Honda agent, who should be able to supply most items ex-stock. Parts cannot be obtained from Honda (UK) Limited direct; all orders must be routed via an approved agent, even if the parts required are not held in stock.

Always quote the engine and frame numbers in full, particularly if parts are required for any of the earlier models.

The frame number is located on the left hand side of the steering head and the engine number is stamped on the upper crankcase, immediately to the rear of the two left hand cylinders. Use only parts of genuine Honda manufacture. Pattern parts are available, some of which originate from Japan, but in many instances they may have an adverse effect on performance and/or reliability. Furthermore the fitting of non-standard parts may invalidate the warranty. Honda do not operate a 'service exchange' scheme.

Some of the more expendable parts such as spark plugs, bulbs, tyres, oils and greases etc., can be obtained from accessory shops and motor factors who have convenient opening hours and can often be found not far from home. It is also possible to obtain parts on a Mail Order basis from a number of specialists who advertise regularly in the motor cycle magazines.

Frame number is stamped into steering head

Engine number is embossed on crankcase

Safety first!

Professional motor mechanics are trained in safe working procedures. However enthusiastic you may be about getting on with the job in hand, do take the time to ensure that your safety is not put at risk. A moment's lack of attention can result in an accident, as can failure to observe certain elementary precautions.

There will always be new ways of having accidents, and the following points do not pretend to be a comprehensive list of all dangers; they are intended rather to make you aware of the risks and to encourage a safety-conscious approach to all work you carry out on your vehicle.

Essential DOs and DON'Ts

DON'T start the engine without first ascertaining that the transmission is in neutral.

DON'T suddenly remove the filler cap from a hot cooling system – cover it with a cloth and release the pressure gradually first, or you may get scalded by escaping coolant.

DON'T attempt to drain oil until you are sure it has cooled sufficiently to avoid scalding you.

DON'T grasp any part of the engine, exhaust or silencer without first ascertaining that it is sufficiently cool to avoid burning you.

DON'T allow brake fluid or antifreeze to contact the machine's paintwork or plastic components.

DON'T syphon toxic liquids such as fuel, brake fluid or antifreeze by mouth, or allow them to remain on your skin.

DON'T inhale dust – it may be injurious to health (see *Asbestos* heading).

DON'T allow any spilt oil or grease to remain on the floor – wipe it up straight away, before someone slips on it.

DON'T use ill-fitting spanners or other tools which may slip and cause injury.

DON'T attempt to lift a heavy component which may be beyond your capability – get assistance.

DON'T rush to finish a job, or take unverified short cuts.

DON'T allow children or animals in or around an unattended vehicle.

DON'T inflate a tyre to a pressure above the recommended maximum. Apart from overstressing the carcase and wheel rim, in extreme cases the tyre may blow off forcibly.

DO ensure that the machine is supported securely at all times. This is especially important when the machine is blocked up to aid wheel or fork removal.

DO take care when attempting to slacken a stubborn nut or bolt. It is generally better to pull on a spanner, rather than push, so that if slippage occurs you fall away from the machine rather than on to it.

DO wear eye protection when using power tools such as drill, sander, bench grinder etc.

DO use a barrier cream on your hands prior to undertaking dirty jobs – it will protect your skin from infection as well as making the dirt easier to remove afterwards; but make sure your hands aren't left slippery. Note that long-term contact with used engine oil can be a health hazard.

DO keep loose clothing (cuffs, tie etc) and long hair well out of the way of moving mechanical parts.

DO remove rings, wristwatch etc, before working on the vehicle – especially the electrical system.

DO keep your work area tidy – it is only too easy to fall over articles left lying around.

DO exercise caution when compressing springs for removal or installation. Ensure that the tension is applied and released in a controlled manner, using suitable tools which preclude the possibility of the spring escaping violently.

DO ensure that any lifting tackle used has a safe working load rating adequate for the job.

DO get someone to check periodically that all is well, when working alone on the vehicle.

DO carry out work in a logical sequence and check that everything is correctly assembled and tightened afterwards.

DO remember that your vehicle's safety affects that of yourself and others. If in doubt on any point, get specialist advice.

IF, in spite of following these precautions, you are unfortunate enough to injure yourself, seek medical attention as soon as possible.

Asbestos

Certain friction, insulating, sealing, and other products – such as brake linings, clutch linings, gaskets, etc – contain asbestos. *Extreme care must be taken to avoid inhalation of dust from such products since it is hazardous to health*. If in doubt, assume that they *do* contain asbestos.

Fire

Remember at all times that petrol (gasoline) is highly flammable. Never smoke, or have any kind of naked flame around, when working on the vehicle. But the risk does not end there – a spark caused by an electrical short-circuit, by two metal surfaces contacting each other, by careless use of tools, or even by static electricity built up in your body under certain conditions, can ignite petrol vapour, which in a confined space is highly explosive.

Always disconnect the battery earth (ground) terminal before working on any part of the fuel or electrical system, and never risk spilling fuel on to a hot engine or exhaust.

It is recommended that a fire extinguisher of a type suitable for fuel and electrical fires is kept handy in the garage or workplace at all times. Never try to extinguish a fuel or electrical fire with water.

Note: *Any reference to a 'torch' appearing in this manual should always be taken to mean a hand-held battery-operated electric lamp or flashlight. It does **not** mean a welding/gas torch or blowlamp.*

Fumes

Certain fumes are highly toxic and can quickly cause unconsciousness and even death if inhaled to any extent. Petrol (gasoline) vapour comes into this category, as do the vapours from certain solvents such as trichloroethylene. Any draining or pouring of such volatile fluids should be done in a well ventilated area.

When using cleaning fluids and solvents, read the instructions carefully. Never use materials from unmarked containers – they may give off poisonous vapours.

Never run the engine of a motor vehicle in an enclosed space such as a garage. Exhaust fumes contain carbon monoxide which is extremely poisonous; if you need to run the engine, always do so in the open air or at least have the rear of the vehicle outside the workplace.

The battery

Never cause a spark, or allow a naked light, near the vehicle's battery. It will normally be giving off a certain amount of hydrogen gas, which is highly explosive.

Always disconnect the battery earth (ground) terminal before working on the fuel or electrical systems.

If possible, loosen the filler plugs or cover when charging the battery from an external source. Do not charge at an excessive rate or the battery may burst.

Take care when topping up and when carrying the battery. The acid electrolyte, even when diluted, is very corrosive and should not be allowed to contact the eyes or skin.

If you ever need to prepare electrolyte yourself, always add the acid slowly to the water, and never the other way round. Protect against splashes by wearing rubber gloves and goggles.

Mains electricity and electrical equipment

When using an electric power tool, inspection light etc, always ensure that the appliance is correctly connected to its plug and that, where necessary, it is properly earthed (grounded). Do not use such appliances in damp conditions and, again, beware of creating a spark or applying excessive heat in the vicinity of fuel or fuel vapour. Also ensure that the appliances meet the relevant national safety standards.

Ignition HT voltage

A severe electric shock can result from touching certain parts of the ignition system, such as the HT leads, when the engine is running or being cranked, particularly if components are damp or the insulation is defective. Where an electronic ignition system is fitted, the HT voltage is much higher and could prove fatal.

Tools and working facilities

The first priority when undertaking maintenance or repair work of any sort on a motorcycle is to have a clean, dry, well-lit working area. Work carried out in peace and quiet in the well-ordered atmosphere of a good workshop will give more satisfaction and much better results than can usually be achieved in poor working conditions. A good workshop must have a clean flat workbench or a solidly constructed table of convenient working height. The workbench or table should be equipped with a vice which has a jaw opening of at least 4 in (100 mm). A set of jaw covers should be made from soft metal such as aluminium alloy or copper, or from wood. These covers will minimise the marking or damaging of soft or delicate components which may be clamped in the vice. Some clean, dry, storage space will be required for tools, lubricants and dismantled components. It will be necessary during a major overhaul to lay out engine/gearbox components for examination and to keep them where they will remain undisturbed for as long as is necessary. To this end it is recommended that a supply of metal or plastic containers of suitable size is collected. A supply of clean, lint-free, rags for cleaning purposes and some newspapers, other rags, or paper towels for mopping up spillages should also be kept. If working on a hard concrete floor note that both the floor and one's knees can be protected from oil spillages and wear by cutting open a large cardboard box and spreading it flat on the floor under the machine or workbench. This also helps to provide some warmth in winter and to prevent the loss of nuts, washers, and other tiny components which have a tendency to disappear when dropped on anything other than a perfectly clean, flat, surface.

Unfortunately, such working conditions are not always available to the home mechanic. When working in poor conditions it is essential to take extra time and care to ensure that the components being worked on are kept scrupulously clean and to ensure that no components or tools are lost or damaged.

A selection of good tools is a fundamental requirement for anyone contemplating the maintenance and repair of a motor vehicle. For the owner who does not possess any, their purchase will prove a considerable expense, offsetting some of the savings made by doing-it-yourself. However, provided that the tools purchased meet the relevant national safety standards and are of good quality, they will last for many years and prove an extremely worthwhile investment.

To help the average owner to decide which tools are needed to carry out the various tasks detailed in this manual, we have compiled three lists of tools under the following headings: *Maintenance and minor repair, Repair and overhaul,* and *Specialized*. The newcomer to practical mechanics should start off with the simpler jobs around the vehicle. Then, as his confidence and experience grow, he can undertake more difficult tasks, buying extra tools as and when they are needed. In this way, a *Maintenance and minor repair* tool kit can be built-up into a *Repair and overhaul* tool kit over a considerable period of time without any major cash outlays. The experienced home mechanic will have a tool kit good enough for most repair and overhaul procedures and will add tools from the specialized category when he feels the expense is justified by the amount of use these tools will be put to.

It is obviously not possible to cover the subject of tools fully here. For those who wish to learn more about tools and their use there is a book entitled *Motorcycle Workshop Practice Manual* (Bk no 1454) available from the publishers of this manual.

As a general rule, it is better to buy the more expensive, good quality tools. Given reasonable use, such tools will last for a very long time, whereas the cheaper, poor quality, item will wear out faster and need to be renewed more often, thus nullifying the original saving. There is also the risk of a poor quality tool breaking while in use, causing personal injury or expensive damage to the component being worked on.

For practically all tools, a tool factor is the best source since he will have a very comprehensive range compared with the average garage or accessory shop. Having said that, accessory shops often offer excellent quality tools at discount prices, so it pays to shop around. There are plenty of tools around at reasonable prices, but always aim to purchase items which meet the relevant national safety standards. If in doubt, seek the advice of the shop proprietor or manager before making a purchase.

The basis of any toolkit is a set of spanners. While open-ended spanners with their slim jaws, are useful for working on awkwardly-positioned nuts, ring spanners have advantages in that they grip the nut far more positively. There is less risk of the spanner slipping off the nut and damaging it, for this reason alone ring spanners are to be preferred. Ideally, the home mechanic should acquire a set of each, but if expense rules this out a set of combination spanners (open-ended at one end and with a ring of the same size at the other) will provide a good compromise. Another item which is so useful it should be considered an essential requirement for any home mechanic is a set of socket spanners. These are available in a variety of drive sizes. It is recommended that the ½-inch drive type is purchased to begin with as although bulkier and more expensive than the ⅜-inch type, the larger size is far more common and will accept a greater variety of torque wrenches, extension pieces and socket sizes. The socket set should comprise sockets of sizes between 8 and 24 mm, a reversible ratchet drive, an extension bar of about 10 inches in length, a spark plug socket with a rubber insert, and a universal joint. Other attachments can be added to the set at a later date.

Maintenance and minor repair tool kit

> Set of spanners 8 – 24 mm
> Set of sockets and attachments
> Spark plug spanner with rubber insert – 10, 12, or 14 mm
> as appropriate
> Adjustable spanner
> C-spanner/pin spanner
> Torque wrench (same size drive as sockets)
> Set of screwdrivers (flat blade)
> Set of screwdrivers (cross-head)
> Set of Allen keys 4 – 10 mm
> Impact screwdriver and bits
> Ball pein hammer – 2 lb
> Hacksaw (junior)
> Self-locking pliers – Mole grips or vice grips
> Pliers – combination
> Pliers – needle nose
> Wire brush (small)
> Soft-bristled brush
> Tyre pump
> Tyre pressure gauge
> Tyre tread depth gauge
> Oil can
> Fine emery cloth
> Funnel (medium size)
> Drip tray
> Grease gun
> Set of feeler gauges
> Brake bleeding kit
> Strobe timing light
> Continuity tester (dry battery and bulb)
> Soldering iron and solder
> Wire stripper or craft knife
> PVC insulating tape
> Assortment of split pins, nuts, bolts, and washers

Repair and overhaul toolkit

The tools in this list are virtually essential for anyone undertaking major repairs to a motorcycle and are additional to the tools listed above. Concerning Torx driver bits, Torx screws are encountered on some of the more modern machines where their use is restricted to fastening certain components inside the engine/gearbox unit. It is therefore recommended that if Torx bits cannot be borrowed from a local dealer, they are purchased individually as the need arises. They are not in regular use in the motor trade and will therefore only be available in specialist tool shops.

> Plastic or rubber soft-faced mallet
> Torx driver bits
> Pliers – electrician's side cutters
> Circlip pliers – internal (straight or right-angled tips are
> available)
> Circlip pliers – external
> Cold chisel
> Centre punch
> Pin punch
> Scriber
> Scraper (made from soft metal such as aluminium
> or copper)
> Soft metal drift
> Steel rule/straight edge
> Assortment of files

> Electric drill and bits
> Wire brush (large)
> Soft wire brush (similar to those used for cleaning suede
> shoes)
> Sheet of plate glass
> Hacksaw (large)
> Valve grinding tool
> Valve grinding compound (coarse and fine)
> Stud extractor set (E-Z out)

Specialized tools

This is not a list of the tools made by the machine's manufacturer to carry out a specific task on a limited range of models. Occasional references are made to such tools in the text of this manual and, in general, an alternative method of carrying out the task without the manufacturer's tool is given where possible. The tools mentioned in this list are those which are not used regularly and are expensive to buy in view of their infrequent use. Where this is the case it may be possible to hire or borrow the tools against a deposit from a local dealer or tool hire shop. An alternative is for a group of friends or a motorcycle club to join in the purchase.

> Valve spring compressor
> Piston ring compressor
> Universal bearing puller
> Cylinder bore honing attachment (for electric drill)
> Micrometer set
> Vernier calipers
> Dial gauge set
> Cylinder compression gauge
> Vacuum gauge set
> Multimeter
> Dwell meter/tachometer

Care and maintenance of tools

Whatever the quality of the tools purchased, they will last much longer if cared for. This means in practice ensuring that a tool is used for its intended purpose; for example screwdrivers should not be used as a substitute for a centre punch, or as chisels. Always remove dirt or grease and any metal particles but remember that a light film of oil will prevent rusting if the tools are infrequently used. The common tools can be kept together in a large box or tray but the more delicate, and more expensive, items should be stored separately where they cannot be damaged. When a tool is damaged or worn out, be sure to renew it immediately. It is false economy to continue to use a worn spanner or screwdriver which may slip and cause expensive damage to the component being worked on.

Fastening systems

Fasteners, basically, are nuts, bolts and screws used to hold two or more parts together. There are a few things to keep in mind when working with fasteners. Almost all of them use a locking device of some type; either a lock washer, lock nut, locking tab or thread adhesive. All threaded fasteners should be clean, straight, have undamaged threads and undamaged corners on the hexagon head where the spanner fits. Develop the habit of replacing all damaged nuts and bolts with new ones.

Rusted nuts and bolts should be treated with a rust penetrating fluid to ease removal and prevent breakage. After applying the rust penetrant, let it 'work' for a few minutes before trying to loosen the nut or bolt. Badly rusted fasteners may have to be chiseled off or removed with a special nut breaker, available at tool shops.

Flat washers and lock washers, when removed from an assembly should always be replaced exactly as removed. Replace any damaged washers with new ones. Always use a flat washer between a lock washer and any soft metal surface (such as aluminium), thin sheet metal or plastic. Special lock nuts can only be used once or twice before they lose their locking ability and must be renewed.

If a bolt or stud breaks off in an assembly, it can be drilled out and removed with a special tool called an E-Z out. Most dealer service departments and motorcycle repair shops can perform this task, as well as others (such as the repair of threaded holes that have been stripped out).

Spanner size comparison

Jaw gap (in)	Spanner size	Jaw gap (in)	Spanner size
0.250	$\frac{1}{4}$ in AF	0.945	24 mm
0.276	7 mm	1.000	1 in AF
0.313	$\frac{5}{16}$ in AF	1.010	$\frac{9}{16}$ in Whitworth; $\frac{5}{8}$ in BSF
0.315	8 mm	1.024	26 mm
0.344	$\frac{11}{32}$ in AF; $\frac{1}{8}$ in Whitworth	1.063	$1\frac{1}{16}$ in AF; 27 mm
0.354	9 mm	1.100	$\frac{5}{8}$ in Whitworth; $\frac{11}{16}$ in BSF
0.375	$\frac{3}{8}$ in AF	1.125	$1\frac{1}{8}$ in AF
0.394	10 mm	1.181	30 mm
0.433	11 mm	1.200	$\frac{11}{16}$ in Whitworth; $\frac{3}{4}$ in BSF
0.438	$\frac{7}{16}$ in AF	1.250	$1\frac{1}{4}$ in AF
0.445	$\frac{3}{16}$ in Whitworth; $\frac{1}{4}$ in BSF	1.260	32 mm
0.472	12 mm	1.300	$\frac{3}{4}$ in Whitworth; $\frac{7}{8}$ in BSF
0.500	$\frac{1}{2}$ in AF	1.313	$1\frac{5}{16}$ in AF
0.512	13 mm	1.390	$\frac{13}{16}$ in Whitworth; $\frac{15}{16}$ in BSF
0.525	$\frac{1}{4}$ in Whitworth; $\frac{5}{16}$ in BSF	1.417	36 mm
0.551	14 mm	1.438	$1\frac{7}{16}$ in AF
0.563	$\frac{9}{16}$ in AF	1.480	$\frac{7}{8}$ in Whitworth; 1 in BSF
0.591	15 mm	1.500	$1\frac{1}{2}$ in AF
0.600	$\frac{5}{16}$ in Whitworth; $\frac{3}{8}$ in BSF	1.575	40 mm; $\frac{15}{16}$ in Whitworth
0.625	$\frac{5}{8}$ in AF	1.614	41 mm
0.630	16 mm	1.625	$1\frac{5}{8}$ in AF
0.669	17 mm	1.670	1 in Whitworth; $1\frac{1}{8}$ in BSF
0.686	$\frac{11}{16}$ in AF	1.688	$1\frac{11}{16}$ in AF
0.709	18 mm	1.811	46 mm
0.710	$\frac{3}{8}$ in Whitworth; $\frac{7}{16}$ in BSF	1.813	$1\frac{13}{16}$ in AF
0.748	19 mm	1.860	$1\frac{1}{8}$ in Whitworth; $1\frac{1}{4}$ in BSF
0.750	$\frac{3}{4}$ in AF	1.875	$1\frac{7}{8}$ in AF
0.813	$\frac{13}{16}$ in AF	1.969	50 mm
0.820	$\frac{7}{16}$ in Whitworth; $\frac{1}{2}$ in BSF	2.000	2 in AF
0.866	22 mm	2.050	$1\frac{1}{4}$ in Whitworth; $1\frac{3}{8}$ in BSF
0.875	$\frac{7}{8}$ in AF	2.165	55 mm
0.920	$\frac{1}{2}$ in Whitworth; $\frac{9}{16}$ in BSF	2.362	60 mm
0.938	$\frac{15}{16}$ in AF		

Standard torque settings

Specific torque settings will be found at the end of the specifications section of each chapter. Where no figure is given, bolts should be secured according to the table below.

Fastener type (thread diameter)	kgf m	lbf ft
5mm bolt or nut	0.45 – 0.6	3.5 – 4.5
6 mm bolt or nut	0.8 – 1.2	6 – 9
8 mm bolt or nut	1.8 – 2.5	13 – 18
10 mm bolt or nut	3.0 – 4.0	22 – 29
12 mm bolt or nut	5.0 – 6.0	36 – 43
5 mm screw	0.35 – 0.5	2.5 – 3.6
6 mm screw	0.7 – 1.1	5 – 8
6 mm flange bolt	1.0 – 1.4	7 – 10
8 mm flange bolt	2.4 – 3.0	17 – 22
10 mm flange bolt	3.0 – 4.0	22 – 29

Choosing and fitting accessories

The range of accessories available to the modern motorcyclist is almost as varied and bewildering as the range of motorcycles. This Section is intended to help the owner in choosing the correct equipment for his needs and to avoid some of the mistakes made by many riders when adding accessories to their machines. It will be evident that the Section can only cover the subject in the most general terms and so it is recommended that the owner, having decided that *he wants to fit, for example, a luggage rack or carrier, seeks the advice* of several local dealers and the owners of similar machines. This will give a good idea of what makes of carrier are easily available, and at what price. Talking to other owners will give some insight into the drawbacks or good points of any one make. A walk round the motorcycles in car parks or outside a dealer will often reveal the same sort of information.

The first priority when choosing accessories is to assess exactly what one needs. It is, for example, pointless to buy a large heavy-duty carrier which is designed to take the weight of fully laden panniers and topbox when all you need is a place to strap on a set of waterproofs and a lunchbox when going to work. Many accessory manufacturers have ranges of equipment to cater for the individual needs of different riders and this point should be borne in mind when looking through a dealer's catalogues. Having decided exactly what is required and the use to which the accessories are going to be put, the owner will need a few hints on what to look for when making the final choice. To this end the Section is now sub-divided to cover the more popular accessories fitted. Note that it is in no way a customizing guide, but merely seeks to outline the practical considerations to be taken into account when adding aftermarket equipment to a motorcycle.

Fairings and windscreens

A fairing is possibly the single, most expensive, aftermarket item to be fitted to any motorcycle and, therefore, requires the most thought before purchase. Fairings can be divided into two main groups: front fork mounted handlebar fairings and windscreens, and frame mounted fairings.

The first group, the front fork mounted fairings, are becoming far more popular than was once the case, as they offer several advantages over the second group. Front fork mounted fairings generally are much easier and quicker to fit, involve less modification to the motorcycle, do *not as* a rule restrict the steering lock, permit a wider selection of handlebar styles to be used, and offer adequate protection for much less money than the frame mounted type. They are also lighter, can be swapped easily between different motorcycles, and are available in a much greater variety of styles. Their main disadvantages are that they do not offer as much weather protection as the frame mounted types, rarely offer any storage space, and, if poorly fitted or naturally incompatible, can have an adverse effect on the stability of the motorocycle.

The second group, the frame mounted fairings, are secured so rigidly to the main frame of the motorcycle that they can offer a substantial amount of protection to motorcycle and rider in the event of a crash. They offer almost complete protection from the weather and, if double-skinned in construction, can provide a great deal of useful storage space. The feeling of peace, quiet and complete relaxation encountered when riding behind a good full fairing has to be experienced to be believed. For this reason full fairings are considered essential by most touring motorcyclists and by many people who ride all year round. The main disadvantages of this type are that fitting can take a long time, often involving removal or modification of standard motorcycle components, they restrict the steering lock and they can add up to about 40 lb to the weight of the machine. They do not usually affect the stability of the machine to any great extent once the front tyre pressure and suspension have been adjusted to compensate for the extra weight, but can be affected by sidewinds.

The first thing to look for when purchasing a fairing is the quality of the fittings. A good fairing will have strong, substantial brackets constructed from heavy-gauge tubing; the brackets must be shaped to fit the frame or forks evenly so that the minimum of stress is imposed on the assembly when it is bolted down. The brackets should be properly painted or finished – a nylon coating being the favourite of the better manufacturers – the nuts and bolts provided should be of the same thread and size standard as is used on the motorcycle and be properly plated. Look also for shakeproof locking nuts or locking washers to ensure that everything remains securely tightened down. The fairing shell is generally made from one of two materials: fibreglass or ABS plastic. Both have their advantages and disadvantages, but the main consideration for the owner is that fibreglass is much easier to repair in the event of damage occurring to the fairing. Whichever material is used, check that it is properly finished inside as well as out, that the edges are protected by beading and that the fairing shell is insulated from vibration by the use of rubber grommets at all mounting points. Also be careful to check that the windscreen is retained by plastic bolts which snap on impact so that the windscreen will break away and not cause personal injury in the event of an accident.

Having purchased your fairing or windscreen, read the manufacturer's fitting instructions very carefully and check that you have all the necessary brackets and fittings. Ensure that the mounting brackets are located correctly and bolted down securely. Note that some manufacturers use hose clamps to retain the mounting brackets; these should be discarded as they are convenient to use but not strong enough for the task. Stronger clamps should be substituted; car exhaust pipe clamps of suitable size would be a good alternative. Ensure that the front forks can turn through the full steering lock available without fouling the fairing. With many types of frame-mounted fairing the handlebars will have to be altered or a different type fitted and the steering lock will be restricted by stops provided with the fittings. Also check that the fairing does not foul the front wheel or mudguard, in any steering position, under full fork compression. Re-route any cables, brake pipes or electrical wiring which may snag on the fairing and take great care to protect all electrical connections, using insulating tape. If the manufacturer's instructions are followed carefully at every stage no serious problems should be encountered. Remember that hydraulic pipes that have been disconnected must be carefully re-tightened and the hydraulic system purged of air bubbles by bleeding.

Two things will become immediately apparent when taking a motorcycle on the road for the first time with a fairing – the first is the tendency to underestimate the road speed because of the lack of wind pressure on the body. This must be very carefully watched until one has grown accustomed to riding behind the fairing. The second thing is the alarming increase in engine noise which is an unfortunate but inevitable by-product of fitting any type of fairing or windscreen, and is caused by normal engine noise being reflected, and in some cases amplified, by the flat surface of the fairing.

Luggage racks or carriers

Carriers are possibly the commonest item to be fitted to modern motorcycles. They vary enormously in size, carrying capacity, and durability. When selecting a carrier, always look for one which is made specifically for your machine and which is bolted on with as few separate brackets as possible. The universal-type carrier, with its mass of brackets and adaptor pieces, will generally prove too weak to be of any real use. A good carrier should bolt to the main frame, generally using the two suspension unit top mountings and a mudguard mounting bolt as attachment points, and have its luggage platform as low and as far forward as possible to minimise the effect of any load on the machine's stability. Look for good quality, heavy gauge tubing, good welding and good finish. Also ensure that the carrier does not prevent opening of the seat, sidepanels or tail compartment, as appropriate. When using a carrier, be very careful not to overload it. Excessive weight placed so high and so far to the rear of any motorcycle will have an adverse effect on the machine's steering and stability.

Luggage

Motorcycle luggage can be grouped under two headings: soft and hard. Both types are available in many sizes and styles and have advantages and disadvantages in use.

Soft luggage is now becoming very popular because of its lower cost and its versatility. Whether in the form of tankbags, panniers, or strap-on bags, soft luggage requires in general no brackets and no modification to the motorcycle. Equipment can be swapped easily from one motorcycle to another and can be fitted and removed in seconds. Awkwardly shaped loads can easily be carried. The disadvantages of soft luggage are that the contents cannot be secure against the casual thief, very little protection is afforded in the event of a crash, and waterproofing is generally poor. Also, in the case of panniers, carrying capacity is restricted to approximately 10 lb, although this amount will vary considerably depending on the manufacturer's recommendation. When purchasing soft luggage, look for good quality material, generally vinyl or nylon, with strong, well-stitched attachment points. It is always useful to have separate pockets, especially on tank bags, for items which will be needed on the journey. When purchasing a tank bag, look for one which has a separate, well-padded, base. This will protect the tank's paintwork and permit easy access to the filler cap at petrol stations.

Hard luggage is confined to two types: panniers, and top boxes or tail trunks. Most hard luggage manufacturers produce matching sets of these items, the basis of which is generally that manufacturer's own heavy-duty luggage rack. Variations on this theme occur in the form of separate frames for the better quality panniers, fixed or quickly-detachable luggage, and in size and carrying capacity. Hard luggage offers a reasonable degree of security against theft and good protection against weather and accident damage. Carrying capacity is greater than that of soft luggage, around 15 – 20 lb in the case of panniers, although top boxes should never be loaded as much as their apparent capacity might imply. A top box should only be used for lightweight items, because one that is heavily laden can have a serious effect on the stability of the machine. When purchasing hard luggage look for the same good points as mentioned under fairings and windscreens, ie good quality mounting brackets and fittings, and well-finished fibreglass or ABS plastic cases. Again as with fairings, always purchase luggage made specifically for your motorcycle, using as few separate brackets as possible, to ensure that everything remains securely bolted in place. When fitting hard luggage, be careful to check that the rear suspension and brake operation will not be impaired in any way and remember that many pannier kits require re-siting of the indicators. Remember also that a non-standard exhaust system may make fitting extremely difficult.

Handlebars

The occupation of fitting alternative types of handlebar is extremely popular with modern motorcyclists, whose motives may vary from the purely practical, wishing to improve the comfort of their machines, to the purely aesthetic, where form is more important than function. Whatever the reason, there are several considerations to be borne in mind when changing the handlebars of your machine. If fitting lower bars, check carefully that the switches and cables do not foul the petrol tank on full lock and that the surplus length of cable, brake pipe, and electrical wiring are smoothly and tidily disposed of. Avoid tight kinks in cable or brake pipes which will produce stiff controls or the premature and disastrous failure of an overstressed component. If necessary, remove the petrol tank and re-route the cable from the engine/gearbox unit upwards, ensuring smooth gentle curves are produced. In extreme cases, it will be necessary to purchase a shorter brake pipe to overcome this problem. In the case of higher handlebars than standard it will almost certainly be necessary to purchase extended cables and brake pipes. Fortunately, many standard motorcycles have a custom version which will be equipped with higher handlebars and, therefore, factory-built extended components will be available from your local dealer. It is not usually necessary to extend electrical wiring, as switch clusters may be used on several different motorcycles, some being custom versions. This point should be borne in mind however when fitting extremely high or wide handlebars.

When fitting different types of handlebar, ensure that the mounting clamps are correctly tightened to the manufacturer's specifications and that cables and wiring, as previously mentioned, have smooth easy runs and do not snag on any part of the motorcycle throughout the full steering lock. Ensure that the fluid level in the front brake master cylinder remains level to avoid any chance of air entering the hydraulic system. Also check that the cables are adjusted correctly and that all handlebar controls operate correctly and can be easily reached when riding.

Crashbars

Crashbars, also known as engine protector bars, engine guards, or case savers, are extremely useful items of equipment which can contribute protection to the machine's structure if a crash occurs. They do not, as has been inferred in the US, prevent the rider from crashing, or necessarily prevent rider injury should a crash occur.

It is recommended that only the smaller, neater, engine protector type of crashbar is considered. This type will offer protection while restricting, as little as is possible, access to the engine and the machine's ground clearance. The crashbars should be designed for use specifically on your machine, and should be constructed of heavy-gauge tubing with strong, integral mounting brackets. Where possible, they should bolt to a strong lug on the frame, usually at the engine mounting bolts.

The alternative type of crashbar is the larger cage type. This type is not recommended in spite of their appearance which promises some protection to the rider as well as to the machine. The larger amount of leverage imposed by the size of this type of crashbar increases the risk of severe frame damage in the event of an accident. This type also decreases the machine's ground clearance and restricts access to the engine. The amount of protection afforded the rider is open to some doubt as the design is based on the premise that the rider will stay in the normally seated position during an accident, and the crash bar structure will not itself fail. Neither result can in any way be guaranteed.

As a general rule, always purchase the best, ie usually the most expensive, set of crashbars you an afford. The investment will be repaid by minimising the amount of damage incurred, should the machine be involved in an accident. Finally, avoid the universal type of crashbar. This should be regarded only as a last resort to be used if no alternative exists. With its usual multitude of separate brackets and spacers, the universal crashbar is far too weak in design and construction to be of any practical value.

Exhaust systems

The fitting of aftermarket exhaust systems is another extremely popular pastime amongst motorcyclists. The usual motive is to gain more performance from the engine but other considerations are to gain more ground clearance, to lose weight from the motorcycle, to obtain a more distinctive exhaust note or to find a cheaper alternative to the manufacturer's original equipment exhaust system. Original equipment exhaust systems often cost more and may well have a relatively short life. It should be noted that it is rare for an aftermarket exhaust system alone to give a noticeable increase in the engine's power output. Modern motorcycles are designed to give the highest power output possible allowing for factors such as quietness, fuel economy, spread of power, and long-term reliability. If there were a magic formula which allowed the exhaust system to produce more power without affecting these other considerations you can be sure that the manufacturers, with their large research and development facilities, would have found it and made use of it. Performance increases of a worthwhile and noticeable nature only come from well-tried and properly matched modifications to the entire engine, from the air filter, through the carburettors, port timing or camshaft and valve design, combustion chamber shape, compression ratio, and the exhaust system. Such modifications are well outside the scope of this manual but interested owners might refer to specialist books available from the publisher of this manual, which go into the subject in great detail.

Whatever your motive for wishing to fit an alternative exhaust system, be sure to seek advice before doing so. Changes to the carburettor jetting will almost certainly be required for which you must consult the exhaust system manufacturer. If he cannot supply adequately specific information it is reasonable to assume that insufficient development work has been carried out, and that particular make should be avoided. Other factors to be borne in mind are whether the exhaust system allows the use of both centre and side stands, whether it allows sufficient access to permit oil and filter changing and whether modifications are necessary to the standard exhaust system. Many two-stroke expansion chamber systems require the use of the standard exhaust pipe; this is all very well if the standard exhaust pipe and silencer are separate units but can cause problems if the two, as with so many modern two-strokes, are a one-piece unit. While the exhaust pipe can be removed easily by means of a hacksaw it is not so easy to refit the original silencer should you at any time wish to return the

machine to standard trim. The same applies to several four-stroke systems.

On the subject of the finish of aftermarket exhausts, avoid black-painted systems unless you enjoy painting. As any trail-bike owner will tell you, rust has a great affinity for black exhausts and re-painting or rust removal becomes a task which must be carried out with monotonous regularity. A bright chrome finish is, as a general rule, a far better proposition as it is much easier to keep clean and to prevent rusting. Although the general finish of aftermarket exhaust systems is not always up to the standard of the original equipment the lower cost of such systems does at least reflect this fact.

When fitting an alternative system always purchase a full set of new exhaust gaskets, to prevent leaks. Fit the exhaust first to the cylinder head or barrel, as appropriate, tightening the retaining nuts or bolts by hand only and then line up the exhaust rear mountings. If the new system is a one-piece unit and the rear mountings do not line up exactly, spacers must be fabricated to take up the difference. Do not force the system into place as the stress thus imposed will rapidly cause cracks and splits to appear. Once all the mountings are loosely fixed, tighten the retaining nuts or bolts securely, being careful not to overtighten them. Where the motorcycle manufacturer's torque settings are available, these should be used. Do not forget to carry out any carburation changes recommended by the exhaust system's manufacturer.

Electrical equipment

The vast range of electrical equipment available to motorcyclists is so large and so diverse that only the most general outline can be given here. Electrical accessories vary from electric ignition kits fitted to replace contact breaker points, to additional lighting at the front and rear, more powerful horns, various instruments and gauges, clocks, anti-theft systems, heated clothing, CB radios, radio-cassette players, and intercom systems, to name but a few of the more popular items of equipment.

As will be evident, it would require a separate manual to cover this subject alone and this section is therefore restricted to outlining a few basic rules which must be borne in mind when fitting electrical equipment. The first consideration is whether your machine's electrical system has enough reserve capacity to cope with the added demand of the accessories you wish to fit. The motorcycle's manufacturer or importer should be able to furnish this sort of information and may also be able to offer advice on uprating the electrical system. Failing this, a good dealer or the accessory manufacturer may be able to help. In some cases, more powerful generator components may be available, perhaps from another motorcycle in the manufacturer's range. The second consideration is the legal requirements in force in your area. The local police may be prepared to help with this point. In the UK for example, there are strict regulations governing the position and use of auxiliary riding lamps and fog lamps.

When fitting electrical equipment always disconnect the battery first to prevent the risk of a short-circuit, and be careful to ensure that all connections are properly made and that they are waterproof. Remember that many electrical accesories are designed primarily for use in cars and that they cannot easily withstand the exposure to vibration and to the weather. Delicate components must be rubber-mounted to insulate them from vibration, and sealed carefully to prevent the entry of rainwater and dirt. Be careful to follow exactly the accessory manufacturer's instructions in conjunction with the wiring diagram at the back of this manual.

Accessories – general

Accessories fitted to your motorcycle will rapidly deteriorate if not cared for. Regular washing and polishing will maintain the finish and will provide an opportunity to check that all mounting bolts and nuts are securely fastened. Any signs of chafing or wear should be watched for, and the cause cured as soon as possible before serious damage occurs.

As a general rule, do not expect the re-sale value of your motorcycle to increase by an amount proportional to the amount of money and effort put into fitting accessories. It is usually the case that an absolutely standard motorcycle will sell more easily at a better price than one that has been modified. If you are in the habit of exchanging your machine for another at frequent intervals, this factor should be borne in mind to avoid loss of money.

Fault diagnosis

Contents

1   Introduction

This Section provides an easy reference-guide to the more common ailments that are likely to afflict your machine. Obviously, the opportunities are almost limitless for faults to occur as a result of obscure failures, and to try and cover all eventualities would require a book. Indeed, a number have been written on the subject.

Successful fault diagnosis is not a mysterious 'black art' but the application of a bit of knowledge combined with a systematic and logical approach to the problem. Approach any fault diagnosis by first accurately identifying the symptom and then checking through the list of possible causes, starting with the simplest or most obvious and progressing in stages to the most complex. Take nothing for granted, but above all apply liberal quantities of common sense.

The main symptom of a fault is given in the text as a major heading below which are listed, as Section headings, the various systems or areas which may contain the fault. Details of each possible cause for a fault and the remedial action to be taken are given, in brief, in the paragraphs below each Section heading. Further information should be sought in the relevant Chapter.

Starter motor problems

2   Starter motor not rotating

Engine stop switch off.

Fuse blown. Check the main fuse located behind the battery side cover.

Battery voltage low. Switching on the headlamp and operating the horn will give a good indication of the charge level. If necessary recharge the battery from an external source.

Neutral gear not selected. Where a neutral indicator switch is fitted.

Faulty neutral indicator switch or clutch interlock switch. Check the switch wiring and switches for correct operation.

Ignition switch defective. Check switch for continuity and connections for security.

Engine stop switch defective. Check switch for continuity in 'Run' position. Fault will be caused by broken, wet or corroded switch contacts. Clean or renew as necessary.

Starter button switch faulty. Check continuity of switch. Faults as for engine stop switch.

Starter relay (solenoid) faulty. If the switch is functioning correctly a pronounced click should be heard when the starter button is depressed. This presupposes that current is flowing to the solenoid when the button is depressed.

Wiring open or shorted. Check first that the battery terminal connections are tight and corrosion free. Follow this by checking that all wiring connections are dry, tight and corrosion free. Check also for frayed or broken wiring. Occasionally a wire may become trapped between two moving components, particularly in the vicinity of the steering head, leading to breakage of the internal core but leaving the softer but more resilient outer cover intact. This can cause mysterious intermittent or total power loss.

Starter motor defective. A badly worn starter motor may cause high current drain from a battery without the motor rotating. If current is found to be reaching the motor, after checking the starter button and starter relay, suspect a damaged motor. The motor should be removed for inspection.

3   Starter motor rotates but engine does not turn over

Starter motor clutch defective. Suspect jammed or worn engagement rollers, plungers and springs.

Damaged starter motor drive train. Inspect and renew component where necessary. Failure in this area is unlikely.

4   Starter motor and clutch function but engine will not turn over

Engine seized. Seizure of the engine is always a result of damage to internal components due to lubrication failure, or component breakage resulting from abuse, neglect or old age. A seizing or partially

seized component may go un-noticed until the engine has cooled down and an attempt is made to restart the engine. Suspect first seizure of the valves, valve gear and the pistons. Instantaneous seizure whilst the engine is running indicates component breakage. In either case major dismantling and inspection will be required.

Engine does not start when turned over

5   No fuel flow to carburettor

No fuel or insufficient fuel in tank.

Fuel tap lever position incorrectly selected.

Float chambers require priming after running dry.

Tank filler cap air vent obstructed. Usually caused by dirt or water. Clean the vent orifice.

Fuel tap or filter blocked. Blockage may be due to accumulation of rust or paint flakes from the tank's inner surface or of foreign matter from contaminated fuel. Remove the tap and clean it and the filter. Look also for water droplets in the fuel.

Fuel line blocked. Blockage of the fuel line is more likely to result from a kink in the line rather than the accumulation of debris.

6   Fuel not reaching cylinder

Float chamber not filling. Caused by float needle or floats sticking in up position. This may occur after the machine has been left standing for an extended length of time allowing the fuel to evaporate. When this occurs a gummy residue is often left which hardens to a varnish-like substance. This condition may be worsened by corrosion and crystaline deposits produced prior to the total evaporation of contaminated fuel. Sticking of the float needle may also be caused by wear. In any case removal of the float chamber will be necessary for inspection and cleaning.

Blockage in starting circuit, slow running circuit or jets. Blockage of these items may be attributable to debris from the fuel tank by-passing the filter system or to gumming up as described in paragraph 1. Water droplets in the fuel will also block jets and passages. The carburettor should be dismantled for cleaning.

Fuel level too low. The fuel level in the float chamber is controlled by float height. The float height may increase with wear or damage but will never reduce, thus a low float height is an inherent rather than developing condition. Check the float height and make any necessary adjustment.

7   Engine flooding

Float valve needle worn or stuck open. A piece of rust or other debris can prevent correct seating of the needle against the valve seat thereby permitting an uncontrolled flow of fuel. Similarly, a worn needle or needle seat will prevent valve closure. Dismantle the carburettor float bowl for cleaning and, if necessary, renewal of the worn components.

Fuel level too high. The fuel level is controlled by the float height which may increase due to wear of the float needle, pivot pin or operating tang. Check the float height, and make any necessary adjustment. A leaking float will cause an increase in fuel level, and thus should be renewed.

Cold starting mechanism. Check the choke (starter mechanism) for correct operation. If the mechanism jams in the 'On' position subsequent starting of a hot engine will be difficult.

Blocked air filter. A badly restricted air filter will cause flooding. Check the filter and clean or renew as required. A collapsed inlet hose will have a similar effect.

8   No spark at plug

Ignition switch not on.

Engine stop switch off.

Fuse blown. Check fuse for ignition circuit. See wiring diagram.

Battery voltage low. The current draw required by a starter motor is sufficiently high that an under-charged battery may not have enough spare capacity to provide power for the ignition circuit during starting.

Starter motor inefficient. A starter motor with worn brushes and a worn or dirty commutator will draw excessive amounts of current causing power starvation in the ignition system. See the preceding paragraph. Starter motor overhaul will be required.

Spark plug failure. Clean the spark plug thoroughly and reset the electrode gap. Refer to the spark plug section and the colour condition guide in Chapter 3. If the spark plug shorts internally or has sustained visible damage to the electrodes, core or ceramic insulator it should be renewed. On rare occasions a plug that appears to spark vigorously will fail to do so when refitted to the engine and subjected to the compression pressure in the cylinder.

Spark plug cap or high tension (HT) lead faulty. Check condition and security. Replace if deterioration is evident.

Spark plug cap loose. Check that the spark plug cap fits securely over the plug and, where fitted, the screwed terminal on the plug end is secure.

Shorting due to moisture. Certain parts of the ignition system are susceptible to shorting when the machine is ridden or parked in wet weather. Check particularly the area from the spark plug cap back to the ignition coil. A water dispersant spray may be used to dry out waterlogged components. Recurrence of the problem can be prevented by using an ignition sealant spray after drying out and cleaning.

Ignition or stop switch shorted. May be caused by water, corrosion or wear. Water dispersant and contact cleaning sprays may be used. If this fails to overcome the problem dismantling and visual inspection of the switches will be required.

Shorting or open circuit in wiring. Failure in any wire connecting any of the ignition components will cause ignition malfunction. Check also that all connections are clean, dry and tight.

Ignition coil failure. Check the coil, referring to Chapter 3.

Failure of ignition pickup (pulser coils) or spark unit. See Chapter 3.

9 Weak spark at plug

Feeble sparking at the plug may be caused by any of the faults mentioned in the preceding Section other than those items in paragraphs 1, 2 and 3. Check first the contact breaker assembly and the spark plug, these being the most likely culprits.

10 Compression low

Spark plug loose. This will be self-evident on inspection, and may be accompanied by a hissing noise when the engine is turned over. Remove the plug and check that the threads in the cylinder head are not damaged. Check also that the plug sealing washer is in good condition.

Cylinder head gasket leaking. This condition is often accompanied by a high pitched squeak from around the cylinder head and oil loss, and may be caused by insufficiently tightened cylinder head fasteners, a warped cylinder head or mechanical failure of the gasket material. Re-torqueing the fasteners to the correct specification may seal the leak in some instances but if damage has occurred this course of action will provide, at best, only a temporary cure.

Valve not seating correctly. The failure of a valve to seat may be caused by insufficient valve clearance, pitting of the valve seat or face, carbon deposits on the valve seat or seizure of the valve stem or valve gear components. Valve spring breakage will also prevent correct valve closure. The valve clearances should be checked first and then, if these are found to be in order, further dismantling will be required to inspect the relevant components for failure.

Cylinder, piston and ring wear. Compression pressure will be lost if any of these components are badly worn. Wear in one component is invariably accompanied by wear in another. A top end overhaul will be required.

Piston rings sticking or broken. Sticking of the piston rings may be caused by seizure due to lack of lubrication or heating as a result of poor carburation or incorrect fuel type. Gumming of the rings may result from lack of use, or carbon deposits in the ring grooves. Broken rings result from over-revving, overheating or general wear. In either case a top-end overhaul will be required.

Engine stalls after starting

11 General causes

Improper cold start mechanism operation. Check that the operating controls function smoothly and, where applicable, are correctly adjusted. A cold engine may not require application of an enriched mixture to start initially but may baulk without choke once firing. Likewise a hot engine may start with an enriched mixture but will stop almost immediately if the choke is inadvertently in operation.

Ignition malfunction. See Section 9, 'Weak spark at plug'.

Carburettor incorrectly adjusted. Maladjustment of the mixture strength or idle speed may cause the engine to stop immediately after starting. See Chapter 2.

Fuel contamination. Check for filter blockage by debris or water which reduces, but does not completely stop, fuel flow or blockage of the slow speed circuit in the carburettor by the same agents. If water is present it can often be seen as droplets in the bottom of the float bowl. Clean the filter and, where water is in evidence, drain and flush the fuel tank and float bowl.

Intake air leak. Check for security of the carburettor mounting and hose connections, and for cracks or splits in the hoses. Check also that the carburettor top is secure and that the vacuum gauge adaptor plug (where fitted) is tight.

Air filter blocked or omitted. A blocked filter will cause an over-rich mixture; the omission of a filter will cause an excessively weak mixture. Both conditions will have a detrimental effect on carburation. Clean or renew the filter as necessary.

Fuel filler cap air vent blocked. Usually caused by dirt or water. Clean the vent orifice.

Poor running at idle and low speed

12 Weak spark at plug or erratic firing

Battery voltage low. In certain conditions low battery charge, especially when coupled with a badly sulphated battery, may result in misfiring. If the battery is in good general condition it should be recharged; an old battery suffering from sulphated plates should be renewed.

Spark plug fouled, faulty or incorrectly adjusted. See Section 8 or refer to Chapter 3.

Spark plug cap or high tension lead shorting. Check the condition of both these items ensuring that they are in good condition and dry and that the cap is fitted correctly.

Spark plug type incorrect. Fit plug of correct type and heat range as given in Specifications. In certain conditions a plug of hotter or colder type may be required for normal running.

Ignition timing incorrect. Check the ignition timing statically and dynamically, ensuring that the advance is functioning correctly.

Faulty ignition coil. Partial failure of the coil internal insulation will diminish the performance of the coil. No repair is possible, a new component must be fitted.

Faulty pickup (pulser) coil or spark unit. The former is the more likely cause of a partial failure. See Chapteer 3 for test procedures.

13 Fuel/air mixture incorrect

Intake air leak. See Section 11.

Mixture strength incorrect. Adjust slow running mixture strength using pilot adjustment screw.

Carburettor synchronisation.

Pilot jet or slow running circuit blocked. The carburettor should be removed and dismantled for thorough cleaning. Blow through all jets and air passages with compressed air to clear obstructions.

Air cleaner clogged or omitted. Clean or fit air cleaner element as necessary. Check also that the element and air filter cover are correctly seated.

Cold start mechanism in operation. Check that the choke has not been left on inadvertently and the operation is correct. Where applicable check the operating cable free play.

Fuel level too high or too low. Check the float height and adjust as necessary. See Section 7.

Fuel tank air vent obstructed. Obstruction usually caused by dirt or water. Clean vent orifice.

Valve clearance incorrect. Check, and if necessary, adjust, the clearances.

14 Compression low

See Section 10.

Acceleration poor

15 General causes

All items as for previous Section.

Timing not advancing. This is caused by a failure of the advance control circuit in the spark unit. The unit should be renewed; repair is not practical.

Sticking throttle vacuum piston.

Brakes binding. Usually caused by maladjustment or partial seizure of the operating mechanism due to poor maintenance. Check brake adjustment (where applicable). A bent wheel spindle or warped brake disc can produce similar symptoms.

Poor running or lack of power at high speeds

16 Weak spark at plug or erratic firing

All items as for Section 12.

HT lead insulation failure. Insulation failure of the HT lead and spark plug cap due to old age or damage can cause shorting when the engine is driven hard. This condition may be less noticeable, or not noticeable at all at lower engine speeds.

17 Fuel/air mixture incorrect

All items as for Section 13, with the exception of items 2 and 4.

Main jet blocked. Debris from contaminated fuel, or from the fuel tank, and water in the fuel can block the main jet. Clean the fuel filter, the float bowl area, and if water is present, flush and refill the fuel tank.

Main jet is the wrong size. The standard carburettor jetting is for sea level atmospheric pressure. For high altitudes, usually above 5000 ft, a smaller main jet will be required.

Jet needle and needle jet worn. These can be renewed individually but should be renewed as a pair. Renewal of both items requires partial dismantling of the carburettor.

Air bleed holes blocked. Dismantle carburettor and use compressed air to blow out all air passages.

Reduced fuel flow. A reduction in the maximum fuel flow from the fuel tank to the carburettor will cause fuel starvation, proportionate to the engine speed. Check for blockages through debris or a kinked fuel line.

Vacuum diaphragm split. Renew.

18 Compression low

See Section 10.

Knocking or pinking

19 General causes

Carbon build-up in combustion chamber. After high mileages have been covered large accumulation of carbon may occur. This may glow red hot and cause premature ignition of the fuel/air mixture, in advance of normal firing by the spark plug. Cylinder head removal will be required to allow inspection and cleaning.

Fuel incorrect. A low grade fuel, or one of poor quality may result in compression induced detonation of the fuel resulting in knocking and pinking noises. Old fuel can cause similar problems. A too highly leaded fuel will reduce detonation but will accelerate deposit formation in the combustion chamber and may lead to early pre-ignition as described in item 1.

Spark plug heat range incorrect. Uncontrolled pre-ignition can result from the use of a spark plug the heat range of which is too hot.

Weak mixture. Overheating of the engine due to a weak mixture can result in pre-ignition occurring where it would not occur when engine temperature was within normal limits. Maladjustment, blocked jets or passages and air leaks can cause this condition.

Overheating

20 Firing incorrect

Spark plug fouled, defective or maladjusted. See Section 6.

Spark plug type incorrect. Refer to the Specifications and ensure that the correct plug type is fitted.

Incorrect ignition timing. Timing that is far too much advanced or far too much retarded will cause overheating. Check the ignition timing is correct and that the advance mechanism is functioning.

21 Fuel/air mixture incorrect

Slow speed mixture strength incorrect. Adjust pilot air screw.

Main jet wrong size. The carburettor is jetted for sea level atmospheric conditions. For high altitudes, usually above 5000 ft, a smaller main jet will be required.

Air filter badly fitted or omitted. Check that the filter element is in place and that it and the air filter box cover are sealing correctly. Any leaks will cause a weak mixture.

Induction air leaks. Check the security of the carburettor mountings and hose connections, and for cracks and splits in the hoses. Check also that the carburettor top is secure and that the vacuum gauge adaptor plug (where fitted) is tight.

Fuel level too low. See Section 6.

Fuel tank filler cap air vent obstructed. Clear blockage.

22 Lubrication inadequate

Engine oil too low. Not only does the oil serve as a lubricant by preventing friction between moving components, but it also acts as a coolant. Check the oil level and replenish.

Engine oil overworked. The lubricating properties of oil are lost slowly during use as a result of changes resulting from heat and also contamination. Always change the oil at the recommended interval.

Engine oil of incorrect viscosity or poor quality. Always use the recommended viscosity and type of oil.

Oil filter and filter by-pass valve blocked. Renew filter and clean the by-pass valve.

23 Miscellaneous causes

Engine fins clogged. A build-up of mud in the cylinder head and cylinder barrel cooling fins will decrease the cooling capabilities of the fins. Clean the fins as required.

Clutch operating problems

24 Clutch slip

No clutch lever play. Adjust clutch lever end play according to the procedure in Chapter 1.

Friction plates worn or warped. Overhaul clutch assembly, replacing plates out of specification.

Steel plates worn or warped. Overhaul clutch assembly, replacing plates out of specification.

Clutch springs broken or wear. Old or heat-damaged (from slipping clutch) springs should be replaced with new ones.

Clutch inner cable snagging. Caused by a frayed cable or kinked outer cable. Replace the cable with a new one. Repair of a frayed cable is not advised.

Clutch release mechanism defective. Worn or damaged parts in the clutch release mechanism could include the shaft, cam, actuating arm or pivot. Replace parts as necessary.

Clutch hub and outer drum worn. Severe indentation by the clutch plate tangs of the channels in the hub and drum will cause snagging of the plates preventing correct engagement. If this damage occurs, renewal of the worn components is required.

Lubricant incorrect. Use of a transmission lubricant other than that specified may allow the plates to slip.

25 Clutch drag

Clutch lever play excessive. Adjust lever at bars or at cable end if necessary.

Clutch plates warped or damaged. This will cause a drag on the clutch, causing the machine to creep. Overhaul clutch assembly.

Clutch spring tension uneven. Usually caused by a sagged or broken spring. Check and replace springs.

Engine oil deteriorated. Badly contaminated engine oil and a heavy deposit of oil sludge and carbon on the plates will cause plate sticking. The oil recommended for this machine is of the detergent type, therefore it is unlikely that this problem will arise unless regular oil changes are neglected.

Engine oil viscosity too high. Drag in the plates will result from the use of an oil with too high a viscosity. In very cold weather clutch drag may occur until the engine has reached operating temperature.

Clutch hub and outer drum worn. Indentation by the clutch plate tangs of the channels in the hub and drum will prevent easy plate disengagement. If the damage is light the affected areas may be dressed with a fine file. More pronounced damage will necessitate renewal of the components.

Clutch housing seized to shaft. Lack of lubrication, severe wear or damage can cause the housing to seize to the shaft. Overhaul of the clutch, and perhaps the transmission, may be necessary to repair damage.

Clutch release mechanism defective. Worn or damaged release mechanism parts can stick and fail to provide leverage. Overhaul clutch cover components.

Loose clutch hub nut. Causes drum and hub misalignment, putting a drag on the engine. Engagement adjustment continually varies. Overhaul clutch assembly.

Gear selection problems

26 Gear lever does not return

Weak or broken centraliser spring. Renew the spring.

Gearchange shaft bent or seized. Distortion of the gearchange shaft often occurs if the machine is dropped heavily on the gear lever. Provided that damage is not severe straightening of the shaft is permissible.

27 Gear selection difficult or impossible

Clutch not disengaging fully. See Section 25.

Gearchange shaft bent. This often occurs if the machine is dropped heavily on the gear lever. Straightening of the shaft is permissible if the damage is not too great.

Gearchange arms, pawls or pins worn or damaged. Wear or breakage of any of these items may cause difficulty in selecting one or more gears. Overhaul the selector mechanism.

Gearchange shaft centraliser spring maladjusted. This is often characterised by difficulties in changing up or down, but rarely in both directions. Adjust the centraliser anchor bolt as described in Chapter 1.

Gearchange arm spring broken. Renew spring.

Gearchange drum stopper cam damaged. Failure, rather than

wear, of these items may jam the drum thereby preventing gearchanging. The damaged items must be renewed.

Selector forks bent or seized. This can be caused by dropping the machine heavily on the gearchange lever or as a result of lack of lubrication. Though rare, bending of a shaft can result from a missed gearchange or false selection at high speed.

Selector fork end and pin wear. Pronounced wear of these items and the grooves in the gearchange drum can lead to imprecise selection and, eventually, no selection. Renewal of the worn components will be required.

Structural failure. Failure of any one component of the selector rod and change mechanism will result in improper or fouled gear selection.

28 Jumping out of gear

Stopper arm assembly worn or damaged. Wear of the roller and the cam with which it locates and breakage of the detent spring can cause imprecise gear selection resulting in jumping out of gear. Renew the damaged components.

Gear pinion dogs worn or damaged. Rounding off the dog edges and the mating recesses in adjacent pinion can lead to jumping out of gear when under load. The gears should be inspected and renewed. Attempting to reprofile the dogs is not recommended.

Selector forks, gearchange drum and pinion grooves worn. Extreme wear of these interconnected items can occur after high mileages especially when lubrication has been neglected. The worn components must be renewed.

Gear pinions, bushes and shafts worn. Renew the worn components.

Bent gearchange shaft. Often caused by dropping the machine on the gear lever.

Gear pinion tooth broken. Chipped teeth are unlikely to cause jumping out of gear once the gear has been selected fully; a tooth which is completely broken off, however, may cause problems in this respect and in any event will cause transmission noise.

29 Overselection

Pawl spring weak or broken. Renew the spring.

Stopper arm spring worn or broken. Renew the spring.

Gearchange arm stop pads worn. Repairs can be made by welding and reprofiling with a file.

Selector limiter claw components (where fitted) worn or damaged. Renew the damaged items.

Abnormal engine noise

30 Knocking or pinking

See Section 19.

31 Piston slap or rattling from cylinder

Cylinder bore/piston clearance excessive. Resulting from wear, partial seizure or improper boring during overhaul. This condition can often be heard as a high, rapid tapping noise when the engine is under little or no load, particularly when power is just beginning to be applied. Reboring to the next correct oversize should be carried out and a new oversize piston fitted.

Connecting rod bent. This can be caused by over-revving, trying to start a very badly flooded engine (resulting in a hydraulic lock in the cylinder) or by earlier mechanical failure such as a dropped valve. Attempts at straightening a bent connecting rod from a high performance engine are not recommended. Careful inspection of the crankshaft should be made before renewing the damaged connecting rod.

Gudgeon pin, piston boss bore or small-end bearing wear or seizure. Excess clearance or partial seizure between normal moving parts of these items can cause continuous or intermittent tapping noises. Rapid wear or seizure is caused by lubrication starvation resulting from an insufficient engine oil level or oilway blockage.

Piston rings worn, broken or sticking. Renew the rings after careful inspection of the piston and bore.

32 Valve noise or tapping from the cylinder head

Valve clearance incorrect. Adjust the clearances with the engine cold.

Valve spring broken or weak. Renew the spring set.

Camshaft or cylinder head worn or damaged. The camshaft lobes are the most highly stressed of all components in the engine and are subject to high wear if lubrication becomes inadequate. The bearing surfaces on the camshaft and cylinder head are also sensitive to a lack of lubrication. Lubrication failure due to blocked oilways can occur, but over-enthusiastic revving before engine warm-up is complete is the usual cause.

Worn camshaft drive components. A rustling noise or light tapping which is not improved by correct re-adjustment of the cam chain tension can be emitted by a worn cam chain or worn sprockets and chain. If uncorrected, subsequent cam chain breakage may cause extensive damage. The worn components must be renewed before wear becomes too far advanced.

33 Other noises

Big-end bearing wear. A pronounced knock from within the crankcase which worsens rapidly is indicative of big-end bearing failure as a result of extreme normal wear or lubrication failure. Remedial action in the form of a bottom end overhaul should be taken; continuing to run the engine will lead to further damage including the possibility of connecting rod breakage.

Main bearing failure. Extreme normal wear or failure of the main bearings is characteristically accompanied by a rumble from the crankcase and vibration felt through the frame and footrests. Renew the worn bearings and carry out a very careful examination of the crankshaft.

Crankshaft excessively out of true. A bent crank may result from over-revving or damage from an upper cylinder component or gearbox failure. Damage can also result from dropping the machine on either crankshaft end. Straightening of the crankshaft is not possible in normal circumstances; a replacement item should be fitted.

Engine mounting loose. Tighten all the engine mounting nuts and bolts.

Cylinder head gasket leaking. The noise most often associated with a leaking head gasket is a high pitched squeaking, although any other noise consistent with gas being forced out under pressure from a small orifice can also be emitted. Gasket leakage is often accompanied by oil seepage from around the mating joint or from the cylinder head holding down bolts and nuts. Leakage into the cam chain tunnel or oil return passages will increase crankcase pressure and may cause oil leakage at joints and oil seals. Also, oil contamination will be accelerated. Leakage results from insufficient or uneven tightening of the cylinder head fasteners, or from random mechanical failure. Retightening to the correct torque figure will, at best, only provide a temporary cure. The gasket should be renewed at the earliest opportunity.

Exhaust system leakage. Popping or crackling in the exhaust system, particularly when it occurs with the engine on the overrun, indicates a poor joint either at the cylinder port or at the exhaust pipe/silencer connection. Failure of the gasket or looseness of the clamp should be looked for.

Abnormal transmission noise

34 Clutch noise

Clutch outer drum/friction plate tang clearance excessive.
Clutch outer drum/thrust washer clearance excessive.
Primary drive gear teeth worn or damaged.
Clutch shock absorber assembly worn or damaged.

35 Transmission noise

Bearing or bushes worn or damaged. Renew the affected components.

Gear pinions worn or chipped. Renew the gear pinions.

Metal chips jammed in gear teeth. This can occur when pieces of metal from any failed component are picked up by a meshing pinion. The condition will lead to rapid bearing wear or early gear failure.

Engine/transmission oil level too low. Top up immediately to prevent damage to gearbox and engine.

Gearchange mechanism worn or damaged. Wear or failure of certain items in the selection and change components can induce mis-selection of gears (see Section 27) where incipient engagement of more than one gear set is promoted. Remedial action, by the overhaul of the gearbox, should be taken without delay.

Loose gearbox chain sprocket. Remove the sprocket and check for impact damage to the splines of the sprocket and shaft. Excessive slack between the splines will promote loosening of the securing nut; renewal of the worn components is required. When retightening the nut ensure that it is tightened fully and that, where fitted, the lock washer is bent up against one flat of the nut.

Chain snagging on cases or cycle parts. A badly worn chain or one that is excessively loose may snag or smack against adjacent components.

Exhaust smokes excessively

36 White/blue smoke (caused by oil burning)

Piston rings worn or broken. Breakage or wear of any ring, but particularly the oil control ring, will allow engine oil past the piston into the combustion chamber. Overhaul the cylinder barrel and piston.

Cylinder cracked, worn or scored. These conditions may be caused by overheating, lack of lubrication, component failure or advanced normal wear. The cylinder barrel should be renewed or rebored and the next oversize piston fitted.

Valve oil seal damaged or worn. This can occur as a result of valve guide failure or old age. The emission of smoke is likely to occur when the throttle is closed rapidly after acceleration, for instance, when changing gear. Renew the valve oil seals and, if necessary, the valve guides.

Valve guides worn. See the preceding paragraph.

Engine oil level too high. This increases the crankcase pressure and allows oil to be forced past the piston rings. Often accompanied by seepage of oil at joints and oil seals.

Cylinder head gasket blown between cam chain tunnel or oil return passage. Renew the cylinder head gasket.

Abnormal crankcase pressure. This may be caused by blocked breather passages or hoses causing back-pressure at high engine revolutions.

37 Black smoke (caused by over-rich mixture)

Air filter element clogged. Clean or renew the element.

Main jet loose or too large. Remove the float chamber to check for tightness of the jet. If the machine is used at high altitudes rejetting will be required to compensate for the lower atmospheric pressure.

Cold start mechanism jammed on. Check that the mechanism works smoothly and correctly and that, where fitted, the operating cable is lubricated and not snagged.

Fuel level too high. The fuel level is controlled by the float height which can increase as a result of wear or damage. Remove the float bowl and check the float height. Check also that floats have not punctured; a punctured float will loose buoyancy and allow an increased fuel level.

Float valve needle stuck open. Caused by dirt or a worn valve. Clean the float chamber or renew the needle and, if necessary, the valve seat.

Oil pressure indicator lamp goes on

38 Engine lubrication system failure

Engine oil defective. Oil pump shaft or locating pin sheared off from ingesting debris or seizing from lack of lubrication (low oil level).

Engine oil screen clogged. Change oil and filter and service pickup screen.

Engine oil level too low. Inspect for leak or other problem causing low oil level and add recommended lubricant.

Engine oil viscosity too low. Very old, thin oil, or an improper weight of oil used in engine. Change to correct lubricant.

Camshaft or journals worn. High wear causing drop in oil pressure. Replace cam and/or head. Abnormal wear could be caused by oil starvation at high rpm from low oil level, improper oil weight or type, or loose oil fitting on upper cylinder oil line.

Crankshaft and/or bearings worn. Same problems as paragraph 5. Overhaul lower end.

Relief valve stuck open. This causes the oil to be dumped back into the sump. Repair or replace.

39 Electrical system failure

Oil pressure switch defective. Check switch according to the procedures in Chapter 6. Replace if defective.

Oil pressure indicator lamp wiring system defective. Check for pinched, shorted, disconnected or damaged wiring.

Poor handling or roadholding

40 Directional instability

Steering head bearing adjustment too tight. This will cause rolling or weaving at low speeds. Re-adjust the bearings.

Steering head bearing worn or damaged. Correct adjustment of the bearing will prove impossible to achieve if wear or damage has occurred. Inconsistent handling will occur including rolling or weaving at low speed and poor directional control at indeterminate higher speeds. The steering head bearing should be dismantled for inspection and renewed if required. Lubrication should also be carried out.

Bearing races pitted or dented. Impact damage caused, perhaps, by an accident or riding over a pot-hole can cause indentation of the bearing, usually in one position. This should be noted as notchiness when the handlebars are turned. Renew and lubricate the bearings.

Steering stem bent. This will occur only if the machine is subjected to a high impact such as hitting a curb or a pot-hole. The lower yoke/stem should be renewed; do not attempt to straighten the stem.

Front or rear tyre pressures too low.

Front or rear tyre worn. General instability, high speed wobbles and skipping over white lines indicates that tyre renewal may be required. Tyre induced problems, in some machine/tyre combinations, can occur even when the tyre in question is by no means fully worn.

Swinging arm or suspension link bearings worn. Difficulties in holding line, particularly when cornering or when changing power settings indicates wear in the swinging arm bearings. The swinging arm should be removed from the machine and the bearings renewed.

Swinging arm flexing. The symptoms given in the preceding paragraph will also occur if the swinging arm fork flexes badly. This can be caused by structural weakness as a result of corrosion, fatigue or impact damage, or because the rear wheel spindle is slack.

Wheel bearings worn. Renew the worn bearings.

Tyres unsuitable for machine. Not all available tyres will suit the characteristics of the frame and suspension, indeed, some tyres or tyre combinations may cause a transformation in the handling characteristics. If handling problems occur immediately after changing to a new tyre type or make, revert to the original tyres to see whether an improvement can be noted. In some instances a change to what are, in fact, suitable tyres may give rise to handling deficiences. In this case a thorough check should be made of all frame and suspension items which affect stability.

41 Steering bias to left or right

Rear wheel out of alignment. Caused by uneven adjustment of chain tensioner adjusters allowing the wheel to be askew in the fork ends. A bent rear wheel spindle will also misalign the wheel in the swinging arm.

Wheels out of alignment. This can be caused by impact damage to the frame, swinging arm, wheel spindles or front forks. Although occasionally a result of material failure or corrosion it is usually as a result of a crash.

Front forks twisted in the steering yokes. A light impact, for instance with a pot-hole or low curb, can twist the fork legs in the steering yokes without causing structural damage to the fork legs or the yokes themselves. Re-alignment can be made by loosening the yoke pinch bolts, wheel spindle and mudguard bolts. Re-align the wheel with the handlebars and tighten the bolts working upwards from the wheel spindle. This action should be carried out only when there is no chance that structural damage has occurred.

42 Handlebar vibrates or oscillates

Tyres worn or out of balance. Either condition, particularly in the front tyre, will promote shaking of the fork assembly and thus the handlebars. A sudden onset of shaking can result if a balance weight is displaced during use.

Tyres badly positioned on the wheel rims. A moulded line on each wall of a tyre is provided to allow visual verification that the tyre is correctly positioned on the rim. A check can be made by rotating the tyre; any misalignment will be immediately obvious.

Wheel rims warped or damaged. Inspect the wheels for runout as described in Chapter 5.

Swinging arm or suspension link bearings worn. Renew the bearings.

Wheel bearings worn. Renew the bearings.

Steering head bearings incorrectly adjusted. Vibration is more likely to result from bearings which are too loose rather than too tight. Re-adjust the bearings.

Loose fork component fasteners. Loose nuts and bolts holding the fork legs, wheel spindle, mudguards or steering stem can promote shaking at the handlebars. Fasteners on running gear such as the forks and suspension should be check tightened occasionally to prevent dangerous looseness of components occurring.

Engine mounting bolts loose. Tighten all fasteners.

43 Poor front fork performance

Air pressure adjustment outside limits. Damping fluid level incorrect. If the fluid level is too low poor suspension control will occur resulting in a general impairment of roadholding and early loss of tyre adhesion when cornering and braking. Too much oil is unlikely to change the fork characteristics unless severe overfilling occurs when the fork action will become stiffer and oil seal failure may occur.

Damping oil viscosity incorrect. The damping action of the fork is directly related to the viscosity of the damping oil. The lighter the oil used, the less will be the damping action imparted. For general use, use the recommended viscosity of oil, changing to a slightly higher or heavier oil only when a change in damping characteristic is required. Overworked oil, or oil contaminated with water which has found its way past the seals, should be renewed to restore the correct damping performance and to prevent bottoming of the forks.

Damping components worn or corroded. Advanced normal wear of the fork internals is unlikely to occur until a very high mileage has been covered. Continual use of the machine with damaged oil seals which allows the ingress of water, or neglect, will lead to rapid corrosion and wear. Dismantle the forks for inspection and overhaul. See Chapter 4.

Weak fork springs. Progressive fatigue of the fork springs, resulting in a reduced spring free length, will occur after extensive use. This condition will promote excessive fork dive under braking, and in its advanced form will reduce the at-rest extended length of the forks and thus the fork geometry. Renewal of the springs as a pair is the only satisfactory course of action.

Bent stanchions or corroded stanchions. Both conditions will prevent correct telescoping of the fork legs, and in an advanced state can cause sticking of the fork in one position. In a mild form corrosion will cause stiction of the fork thereby increasing the time the suspension takes to react to an uneven road surface. Bent fork stanchions should be attended to immediately because they indicate that impact damage has occurred, and there is a danger that the forks will fail with disastrous consequences.

44 Front fork judder when braking (see also Section 56)

Wear between the fork stanchions and the fork legs. Renewal of the affected components is required.

Slack steering head bearings. Re-adjust the bearings.

Warped brake disc. If irregular braking action occurs fork judder can be induced in what are normally serviceable forks. Renew the damaged brake components.

45 Poor rear suspension performance

Air pressure adjustment outside limits. Rear suspension unit damper worn out or leaking. The damping performance of most rear suspension units falls off with age. This is a gradual process, and thus may not be immediately obvious. Indications of poor damping include hopping of the rear end when cornering or braking, and a general loss of positive stability. See Chapter 4.

Weak rear spring. If the suspension unit springs fatigue, they will promote excessive pitching of the machine and reduce the ground clearance when cornering. Although replacement springs are available separately from the rear suspension damper unit it is probable that if spring fatigue has occurred the damper units will also require renewal.

Swinging arm flexing or bearings worn. See Sections 40 and 41.

Bent suspension unit damper rod. This is likely to occur only if the machine is dropped or if seizure of the piston occurs. If either happens the suspension units should be renewed as a pair.

Abnormal frame and suspension noise

46 Front end noise

Oil level low or too thin. This can cause a 'spurting' sound and is usually accompanied by irregular fork action.

Spring weak or broken. Makes a clicking or scraping sound. Fork oil will have a lot of metal particles in it.

Steering head bearings loose or damaged. Clicks when braking. Check, adjust or replace.

Fork clamps loose. Make sure all fork clamp pinch bolts are tight.

Fork stanchion bent. Good possibility if machine has been dropped. Repair or replace tube.

47 Rear suspension noise

Fluid level too low. Leakage of a suspension unit, usually evident by oil on the outer surfaces, can cause a spurting noise. The suspension unit should be renewed.

Defective rear suspension unit with internal damage. Renew the suspension unit.

Brake problems

48 Brakes are spongy or ineffective

Air in brake circuit. This is only likely to happen in service due to neglect in checking the fluid level or because a leak has developed. The problem should be identified and the brake system bled of air.

Pad worn. Check the pad wear against the wear lines provided and renew the pads if necessary.

Contaminated pads. Cleaning pads which have been contaminated with oil, grease or brake fluid is unlikely to prove successful; the pads should be renewed.

Pads glazed. This is usually caused by overheating. The surface of the pads may be roughened using glass-paper or a fine file.

Brake fluid deterioration. A brake which on initial operation is firm but rapidly becomes spongy in use may be failing due to water contamination of the fluid. The fluid should be drained and then the system refilled and bled.

Master cylinder seal failure. Wear or damage of master cylinder internal parts will prevent pressurisation of the brake fluid. Overhaul the master cylinder unit.

Caliper seal failure. This will almost certainly be obvious by loss of fluid, a lowering of fluid in the master cylinder reservoir and contamination of the brake pads and caliper. Overhaul the caliper assembly.

Brake lever or pedal improperly adjusted. Adjust the clearance between the lever end and master cylinder plunger to take up lost motion, as recommended in Routine maintenance.

49 Brakes drag

Disc warped. The disc must be renewed.

Caliper piston, caliper or pads corroded. The brake caliper assembly is vulnerable to corrosion due to water and dirt, and unless cleaned at regular intervals and lubricated in the recommended manner, will become sticky in operation.

Piston seal deteriorated. The seal is designed to return the piston in the caliper to the retracted position when the brake is released. Wear or old age can affect this function. The caliper should be overhauled if this occurs.

Brake pad damaged. Pad material separating from the backing plate due to wear or faulty manufacture. Renew the pads. Faulty installation of a pad also will cause dragging.

Wheel spindle bent. The spindle may be straightened if no structural damage has occurred.

Brake lever or pedal not returning. Check that the lever or pedal works smoothly throughout its operating range and does not snag on any adjacent cycle parts. Lubricate the pivot if necessary.

Twisted caliper support bracket. This is likely to occur only after impact in an accident. No attempt should be made to re-align the caliper; the bracket should be renewed.

50 Brake lever or pedal pulsates in operation

Disc warped or irregularly worn. The disc must be renewed.

Wheel spindle bent. The spindle may be straightened provided no structural damage has occurred.

51 Disc brake noise

Brake squeal. This can be caused by the omission or incorrect installation of the anti-squeal shim fitted to the rear of one pad. The arrow on the shim should face the direction of wheel normal rotation. Squealing can also be caused by dust on the pads, usually in combination with glazed pads, or other contamination from oil, grease, brake fluid or corrosion. Persistent squealing which cannot be traced to any of the normal causes can often be cured by applying a thin layer of high temperature silicone grease to the rear of the pads. Make absolutely certain that no grease is allowed to contaminate the braking surface of the pads.

Glazed pads. This is usually caused by high temperatures or contamination. The pad surfaces may be roughened using glass-paper or a fine file. If this approach does not effect a cure the pads should be renewed.

Disc warped. This can cause a chattering, clicking or intermittent squeal and is usually accompanied by a pulsating brake lever or pedal or uneven braking. The disc must be renewed.

Brake pads fitted incorrectly or undersize. Longitudinal play in the pads due to omission of the locating springs (where fitted) or because pads of the wrong size have been fitted will cause a single tapping noise every time the brake is operated. Inspect the pads for correct installation and security.

52 Brake induced fork judder

Worn front fork stanchions and legs, or worn or badly adjusted steering head bearings. These conditions, combined with uneven or pulsating braking as described in Section 50 will induce more or less judder when the brakes are applied, dependent on the degree of wear and poor brake operation. Attention should be given to both areas of malfunction. See the relevant Sections.

Electrical problems

53 Battery dead or weak

Battery faulty. Battery life should not be expected to exceed 3 to 4 years, particularly where a starter motor is used regularly. Gradual sulphation of the plates and sediment deposits will reduce the battery performance. Plate and insulator damage can often occur as a result of vibration. Complete power failure, or intermittent failure, may be due to a broken battery terminal. Lack of electrolyte will prevent the battery maintaining charge.

Battery leads making poor contact. Remove the battery leads and clean them and the terminals, removing all traces of corrosion and tarnish. Reconnect the leads and apply a coating of petroleum jelly to the terminals.

Load excessive. If additional items such as spot lamps, are fitted, which increase the total electrical load above the maximum alternator output, the battery will fail to maintain full charge. Reduce the electrical load to suit the electrical capacity.

Regulator/rectifier failure.

Alternator generating coils open-circuit or shorted.

Charging circuit shorting or open circuit. This may be caused by frayed or broken wiring, dirty connectors or a faulty ignition switch. The system should be tested in a logical manner. See Section 56.

54 Battery overcharged

Rectifier/regulator faulty. Overcharging is indicated if the battery becomes hot or it is noticed that the electrolyte level falls repeatedly between checks. In extreme cases the battery will boil causing corrosive gases and electrolyte to be emitted through the vent pipes.

Battery wrongly matched to the electrical circuit. Ensure that the specified battery is fitted to the machine.

55 Total electrical failure

Fuse blown. Check the main fuse. If a fault has occurred, it must be rectified before a new fuse is fitted.

Battery faulty. See Section 53.

Earth failure. Check that the frame main earth strap from the battery is securely affixed to the frame and is making a good contact.

Ignition switch or power circuit failure. Check for current flow through the battery positive lead (red) to the ignition switch. Check the ignition switch for continuity.

56 Circuit failure

Cable failure. Refer to the machine's wiring diagram and check the circuit for continuity. Open circuits are a result of loose or corroded connections, either at terminals or in-line connectors, or because of broken wires. Occasionally, the core of a wire will break without there being any apparent damage to the outer plastic cover.

Switch failure. All switches may be checked for continuity in each switch position, after referring to the switch position boxes incorporated in the wiring diagram for the machine. Switch failure may be a result of mechanical breakage, corrosion or water.

Fuse blown. Refer to the wiring diagram to check whether or not a circuit fuse is fitted. Replace the fuse, if blown, only after the fault has been identified and rectified.

57 Bulbs blowing repeatedly

Vibration failure. This is often an inherent fault related to the natural vibration characteristics of the engine and frame and is, thus, difficult to resolve. Modifications of the lamp mounting, to change the damping characteristics may help.

Intermittent earth. Repeated failure of one bulb, particularly where the bulb is fed directly from the generator, indicates that a poor earth exists somewhere in the circuit. Check that a good contact is available at each earthing point in the circuit.

Reduced voltage. Where a quartz-halogen bulb is fitted the voltage to the bulb should be maintained or early failure of the bulb will occur. Do not overload the system with additional electrical equipment in excess of the system's power capacity and ensure that all circuit connections are maintained clean and tight.

HONDA CBX 550

Check list

Weekly or every 200 miles (320 km)

1　Inspect the tyres for damage and low pressures
2　Check the battery electrolyte level
3　Check the engine oil level
4　Check the operation of the electrical system
5　Check the level of hydraulic fluid in the brake master cylinder reservoirs

Monthly or every 600 miles (1000 km)

1　Lubricate and if necessary adjust the final drive chain
2　Check the operation of the suspension and check around the machine for loose fasteners and fittings

Three monthly or every 2000 miles (3200 km)

1　Change the engine oil

Six monthly or every 4000 miles (6400 km)

1　Examine the fuel system hoses and check carburettor adjustment
2　Remove and clean the air filter element
3　Drain the crankcase breather tube
4　Clean and adjust the spark plugs
5　Adjust the valve clearances
6　Renew the oil filter
7　Inspect the brake hydraulic systems for leakage or damage, and check the degree of pad wear
8　Adjust the clutch
9　Check the operation of the electrical system and inspect the components for damage and corrosion
10　Check and lubricate the stand pivots
11　Check the suspension and steering
12　Check the tightness of all bolts and fasteners
13　Check the condition of the wheels

Adjustment data

Valve clearances	
Inlet and exhaust	0.10 – 0.14 mm (0.0039 – 0.0055 in)

Spark plug type	NGK DR8ES or ND X27ESRU
Spark plug gap	0.6 – 0.7 mm (0.024 – 0.028 in)

Ignition timing	
Retarded	15° BTDC ± 1550 ± 200 rpm
Advanced	37° BTDC ± 3000 ± 250 rpm

Idle speed	1200 ± 100 rpm

Tyre pressures	Front	Rear
Solo	32 psi	32 psi
	(2.25 kg/cm²)	(2.25 kg/cm²)
Pillion	32 psi	40 psi
	(2.25 kg/cm²)	(2.80 kg/cm²)

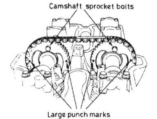

Camshaft sprocket bolts

Large punch marks

Camshaft position for valve clearance check

Recommended lubricants

Component	Quantity	Type/viscosity
① Engine/gearbox	2.3 lit (4.0 Imp pt)	SAE 10W/40
② Final drive chain	As required	SAE 80 or 90 gear oil or aerosol lubricant suitable for sealed chains
③ Front forks		Automatic transmission fluid (ATF)
Right-hand leg	292.5 – 297.5 cc	
Left-hand leg	302.5 – 305.5 cc	
④ Front and rear brake	As required	SAE J1703
⑤ Wheel bearings	As required	High melting point grease
⑥ Steering head bearings	As required	General purpose grease
⑦ Swinging arm bearings	As required	High melting point grease
⑧ Pivot points	As required	General purpose grease
⑨ Control cables	As required	Light machine oil

ROUTINE MAINTENANCE GUIDE

Routine maintenance

Periodical routine maintenance is essential to keep the motorcycle in a peak and safe condition. Routine maintenance also saves money because it provides the opportunity to detect and remedy a fault before it develops further and causes more damage. Maintenance should be undertaken on either a calendar or mileage basis depending on whichever comes sooner. The period between maintenance tasks serves only as a guide since there are many variables eg: age of machine, riding technique and adverse conditions.

The maintenance instructions are generally those recommended by the manufacturer but are supplemented by additional tasks which, through practical experience, the author recommends should be carried out at intervals suggested. The additional tasks are primarily of a preventative nature, which will assist in eliminating unexpected failure of a component or system, due to wear and tear, and increase safety margins when riding.

All the maintenance tasks are described in detail together with the procedures required for accomplishing them. If necessary, more general information on each topic can be found in the relevant Chapter within the main text.

Although no special tools are required for routine maintenance, a good selection of general workshop tools is essential. Included in the tools must be a range of metric ring or combination spanners, a selection of crosshead screwdrivers, and two pairs of circlip pliers, one external opening and the other internal opening. Additionally, owing to the extreme tightness of most casing screws on Japanese machines, an impact screwdriver, together with a choice of large or small cross-head screws bits, is absolutely indispensable. This is particularly so if the engine has not been dismantled since leaving the factory.

Weekly, or every 200 miles (320 km)

---

1 Tyres

Check the tyre pressures. Always check the pressure when the tyres are cold as the heat generated when the machine has been ridden can increase the pressures by as much as 8 psi, giving a totally inaccurate reading. Variations in pressure of as little as 2 psi may alter certain handling characteristics. It is therefore recommended that whatever type of pressure gauge is used, it should be checked occasionally to ensure accurate readings. Do not put absolute faith in 'free air' gauges at garages or petrol stations. They have been known to be in error.

Inspect the tyre treads for cracking or evidence that the outer rubber is leaving the inner cover. Also check the tyre walls for splitting or perishing. Carefully inspect the treads for stones, flints or shrapnel which may have become embedded and be slowly working their way towards the inner surface. Remove such objects with an awl or a small screwdriver.

Tyre pressures – cold

	Solo	With Pillion
Front	32 psi	32 psi
Rear	32 psi	40 psi

2 Battery: examination maintenance and charging

The transparent plastic case of the battery permits the upper and lower levels of the electrolyte to be observed without disturbing the battery by removing the right-hand side cover. Maintenance is normally limited to keeping the electrolyte level between the prescribed upper and lower limits and making sure that the vent tube is not blocked. The lead plates and their separators are also visible through the transparent case, a further guide to the general condition of the battery.

Unless acid is spilt, as may occur if the machine falls over, the electrolyte should always be topped up with distilled water to restore the correct level. If acid is spilt onto any part of the machine, it should be neutralised with an alkali such as washing soda or baking powder and washed away with plenty of water, otherwise serious corrosion will occur. Top up with sulphuric acid of the correct specific gravity (1.260 to 1.280) only when spillage has occurred. Check that the vent pipe is well clear of the frame or any of the other cycle parts.

It is seldom practicable to repair a cracked battery case because the acid present in the joint will prevent the formation of an effective

Battery electrolyte level is visible through translucent case

seal. It is always best to renew a cracked battery, especially in view of the corrosion which will be caused if the acid continues to leak.

If the machine is not used for a period, it is advisable to remove the battery and give it a refresher charge every six weeks or so from a charger. If the battery is permitted to discharge completely, the plates will sulphate and render the battery useless.

Occasionally, check the condition of the battery terminals to ensure that corrosion is not taking place and that the electrical connections are tight. If corrosion has occurred, it should be cleaned away by scraping with a knife and then using emery cloth to remove the final traces. Remake the electrical connections whilst the joint is still clean, then smear the assembly with petroleum jelly (NOT grease) to prevent recurrence of the corrosion. Badly corroded connections can have a high electrical resistance and may give the impression of a complete battery failure.

The normal charging rate for batteries of up to 14 amp hour capacity is $1\frac{1}{2}$ amps. It is permissible to charge at a more rapid rate in an emergency but this shortens the life of the battery, and should be avoided. Always remove the vent caps when recharging a battery, otherwise the gas created within the battery when charging takes place may explode and burst the case with disastrous consequences.

3 Engine oil

Check the engine oil level by means of the dipstick incorporated in

the filler plug which screws into the left-hand side of the crankcase. When taking the reading do not screw the plug into the casing; allow it to rest on the rim of the filler orifice. Replenish the engine oil with oil of the specified grade to the maximum level on the dipstick.

4 Electrical system

Check that the various bulbs are functioning properly, paying particular attention to the rear lamp. It is possible that one of the rear lamp or brake lamp filaments has failed but gone unnoticed. Check that the indicators and horn operate normally. Clean all lenses. If any of the fuses has blown recently, check that the source of the problem has been resolved and that the spare fuse has been renewed.

5 Brake fluid

Check the hydraulic fluid level in the front brake master cylinder reservoir. Before removing the reservoir cap and diaphragm place the handlebars in such a position that the reservoir is approximately vertical. This will prevent spillage. The fluid should lie between the upper and lower lines on the reservoir body. Replenish, if necessary, with hydraulic brake fluid of the correct specification, which is DOT 3 (USA) or SAE-J1703. If the level of fluid in either of the reservoirs is excessively low, check the pads for wear. If the pads are not worn, suspect a fluid leakage in the system. This must be rectified immediately. In the case of the rear disc brake, check the fluid level as

Oil level must lie between limits of hatched area

Front brake reservoir level must be above 'lower' mark

Ensure bars are level to avoid spillage when cap is removed

Rear reservoir has screwed cap. Note level marks

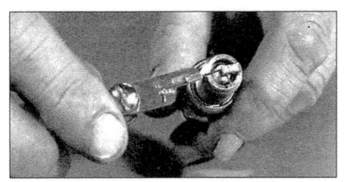

Electrode gap check - use a wire type gauge for best results

Electrode gap adjustment - bend the side electrode using the correct tool

Normal condition - A brown, tan or grey firing end indicates that the engine is in good condition and that the plug type is correct

Ash deposits - Light brown deposits encrusted on the electrodes and insulator, leading to misfire and hesitation. Caused by excessive amounts of oil in the combustion chamber or poor quality fuel/oil

Carbon fouling - Dry, black sooty deposits leading to misfire and weak spark. Caused by an over-rich fuel/air mixture, faulty choke operation or blocked air filter

Oil fouling - Wet oily deposits leading to misfire and weak spark. Caused by oil leakage past piston rings or valve guides (4-stroke engine), or excess lubricant (2-stroke engine)

Overheating - A blistered white insulator and glazed electrodes. Caused by ignition system fault, incorrect fuel, or cooling system fault

Worn plug - Worn electrodes will cause poor starting in damp or cold weather and will also waste fuel

described above. The reservoir is located behind the right-hand side panel.

Monthly or every 600 miles (1000 km)

Complete all the checks listed under the previous maintenance interval heading and then carry out the following.

1 Final drive chain: lubrication and adjustment

The final drive chain is of the endless type, having no joining link in an effort to eliminate any tendency towards breakage. The rollers are equipped with an O-ring at each end which seals the lubricant inside and prevents the ingress of water or abrasive grit. It should not, however, be supposed that the need for lubrication is lessened. On the contrary, frequent but sparse lubrication is essential to minimise wear between the chain and sprockets. Honda recommend the use of SAE 80 or 90 gear oil as a final drive chain lubricant, and warn against the use of aerosol chain lubricants, which they claim can damage the O rings. In practice it will be found that gear oils will quickly become flung off the rotating chain, whilst the sticky aerosol lubricants are designed to adhere to the chain for greater protection. **Note:** To avoid

any risk of O-ring damage, use only those aerosol lubricants, which are designed specifically for use with sealed chains. The propellent or solvent content of other types will cause the O-rings to disintegrate, thus destroying the chain.

In particularly adverse weather conditions, or when touring, lubrication should be undertaken more frequently.

A final word of caution; the importance of chain lubrication cannot be overstressed in view of the cost of replacement, and the fact that a considerable amount of dismantling work, including swinging arm removal, will need to be undertaken should replacement be necessary.

Chain adjustment can be checked with the machine on its centre stand and neutral selected. There should be 20 – 25 mm (0.8 – 1.0 in) free play at the centre of the lower run. It is best to check the chain tension at several points, because wear will rarely be equal at all points. Take the reading at the tightest point. Note that if chain free play reaches 50 mm (2.0 in) or more the chain may contact and damage the frame.

To adjust the chain, slacken the rear wheel spindle nut and the adjuster drawbolt locknuts. Turn each adjuster by an equal amount to maintain wheel alignment. This can be checked by ensuring that the alignment notches on each adjuster are in a similar position in relation to the fork ends. Recheck the adjustment, then secure the wheel spindle (8.5 – 10.5 kgf m, 61 – 67 lbf ft) and the locknuts.

If the chain has stretched to the point where the red zone of the label on the adjusters is in line with the fork end, and free play exceeds 20 mm ($\frac{3}{4}$ in), it must be renewed.

To renew the chain it will be necessary to remove the rear wheel (Chapter 5, Section 4) and the swinging arm (Chapter 4, Section 7). A new chain must never be fitted to worn sprockets, so these should be removed and checked at the same time. In the case of the front sprocket, this is released by removing the sprocket cover, and releasing the two retainer bolts and the retainer. In the case of the rear sprocket, slacken the six sprocket retaining nuts and lift the sprocket away from the cush drive hub. Replace either sprocket if it shows signs of hooking on the loaded tooth faces, or if chipping or breakage of the teeth is discovered.

When fitting the rear sprocket, pull off the cush drive hub and check the condition of the cush drive rubbers. If these show signs of deterioration they should be renewed to avoid harshness in the transmission.

2 Safety check

Give the machine a close visual inspection, securing any loose fasteners or fittings. Check the condition of all control cables and the exposed sections of the wiring harness, and re-route these where chafing is evident. Check that the steering and suspension operate smoothly and evenly, and that the brakes operate normally with no signs of sponginess. Check and lubricate the stand pivots.

Use gear oil or aerosol lubricant suitable for O-ring chains

Slacken wheel spindle and move adjusters by similar amount

Renew chain if 'Replace' mark aligns with end of fork

Remove cush drive hub from wheel ...

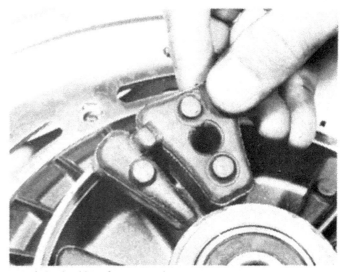
... and check rubbers for wear or damage

Funnel facilitates topping up of crankcase oil

Adjust throttle cable free play at the lower adjuster

Three monthly, or every 2000 miles (3200 km)

Carry out the operations listed under the monthly/600 mile heading, then complete the following.

1 Engine oil: changing

With the engine at full operating temperature, place the machine on its centre stand, remove the crankcase drain plug and allow the oil to drain into a suitable bowl or container. Clean the drain plug and threads, then refit the plug, tightening it to 3.5 – 4.0 kgf m (25 – 30 lbf ft). Top up with new SAE10W/40 motor oil or an approved equivalent (see recommended lubricants). Run the engine for several minutes, then stop it. Check for oil leaks and then check that the oil level is correct before refitting the filler plug.

Six monthly, or every 4000 miles (6400 km)

1 Carburettors and fuel system: checking and adjustment

Check during a normal journey that the idle speed is at its normal setting of 1200 ± 100 rpm, and make any necessary alterations using the central throttle stop control. If uneven or erratic running is evident, and persists after the rest of the six monthly service has been completed, check that the carburettors are correctly synchronised as described in Section 8 of Chapter 2.

Examine the fuel and vacuum hoses, and renew them if any sign of splitting is noted. Check the condition of the throttle cables, renewing them if kinked or frayed. Check that the cables are adjusted to give 2 – 6 mm free play measured at the edge of the twist grip flange. Start the engine and check that the idle speed remains constant when the handlebar is turned from lock-to-lock.

2 Air filter element: cleaning

Unlock and open the dualseat to gain access to the air filter casing. Remove the three screws which retain the cover and lift it away. Grasp the end of the element retaining spring and pull it up to free the element.

Tap the element on a hard surface to dislodge loose dust, then clean it by blowing it through from the inside with compressed air. Examine the element for holes, splits or contamination, and renew it if these are discovered. Refit the element by reversing the removal sequence.

3 Crankcase breather: draining

The crankcase breather incorporates a drain tube which collects emulsified oil vapour during use. The tube is located just to the rear of

Remove air filter casing lid and ...

... pull out spring wire retainer ...

... to allow removal of filter element

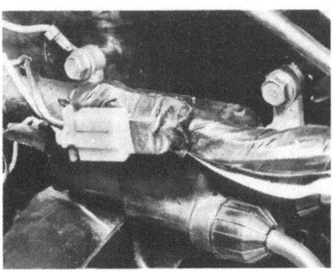

Coil assembly is secured to frame by two bolts

the crankcase and is closed by a plug. Remove the plug and drain off the emulsion into a drip tray, then refit the plug.

4 Spark plugs: cleaning and adjustment

Remove, clean and adjust the sparking plugs. Carbon and other deposits can be removed, using a wire brush, and emery paper or a file used to clean the electrodes prior to adjusting the gaps. Probably the best method of sparking plug cleaning is by having them shot blasted in a special machine. This type of machine is used by most garages. If the outer electrode of a plug is excessively worn (indicated by a step in the underside) the plug should be renewed. Adjust the points gap on each plug by bending the outer electrode only, so that the gap is within the range 0.6 – 0.7 mm (0.024 – 0.028 in). Before replacing the plugs, smear the threads with graphited grease; this will aid subsequent removal. If replacement plugs are required, the correct types are listed at the end of this Section.

5 Valve clearances: adjustment

This operation must only be undertaken when the engine is cold, preferably after it has been left overnight. As a preliminary, lean the motorcycle as far as possible to the left and then right to drain excess oil from the camshaft oil pockets, then place it on its centre stand.

Remove both side panels, open the dualseat and remove the fuel tank (see Chapter 2, Section 2). Pull off the plug caps, then release the

coil mounting bolts and lodge the coil assembly on top of the frame. Release the four bolts which retain the cylinder head cover and lift it away. Remove the left-hand crankshaft end cover to expose the alternator rotor.

Each valve clearance is checked using feeler gauges between the valve stem and adjuster when the relevant cam lobe is positioned opposite the rocker. Honda recommend that two sets of feeler gauges are used to carry out the check. This ensures that the rocker cannot twist and give an inaccurate reading. When adjusted there should be an equal amount of drag on each feeler gauge. To minimise the amount of crankshaft turning required, adopt the following sequence.

Align the index marks (large punch marks) on each camshaft sprocket with the gasket face, then check the inlet valve clearances of those valves where the cam lobes are positioned away from the rockers. Rotate the crankshaft through 180° and check the remaining inlet valves.

Align the camshaft sprocket bolt heads with the cylinder head gasket face, and repeat the above sequence on the exhaust valves.

If the clearances do not correspond with the specified 0.10 – 0.14 mm (0.0039 – 0.0055 in), slacken the adjuster locknut and turn the adjuster in or out until the feeler gauge is a light sliding fit in the gap. Tighten the locknut and re-check the clearance.

During reassembly, check that the cylinder head cover gasket is in sound condition, and apply a little RTV sealant around the extended area of the gasket at the camshaft ends.

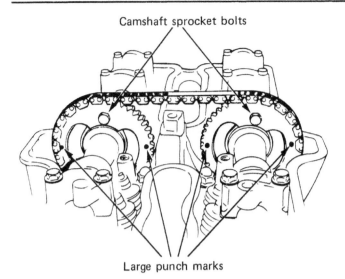

Camshaft sprocket bolts

Large punch marks

Camshaft position for valve clearance check

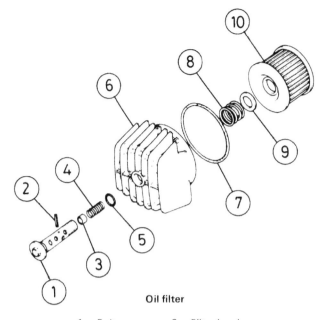

Oil filter

1	Bolt	6	Filter bowl
2	Pin	7	O-ring
3	Collar	8	Spring
4	Spring	9	Washer
5	O-ring	10	Filter element

Measure valve clearances using feeler gauges

6 Oil filter: renewal

The oil filter element must be renewed at each alternate oil change. Start by draining the oil as described under the 3 monthly/2000 mile heading, then place a clean tray beneath the filter housing and remove the filter bolt. Lift away the assembly and discard the element. Clean the housing and crankcase area carefully, then install a new filter element. Check that the O-rings on the filter bolt and around the edge of the housing are in good condition, then refit the assembly, tightening the bolt to 2.8 – 3.2 kgf m (20 – 23 lbf ft) only. Add 2.3 litre (6.0 Imp pint) of SAE10W/40 motor oil or an approved equivalent (see recommended lubricants) having checked that the drain plug has been refitted. Run the engine for several minutes then switch off, check for oil leaks and check the oil level.

7 Braking system: inspection and pad renewal

Check the fluid level in the front and rear reservoirs, and give the hydraulic system a close visual inspection for signs of leakage or damage. **Do not** top up until the pads have been checked for wear.

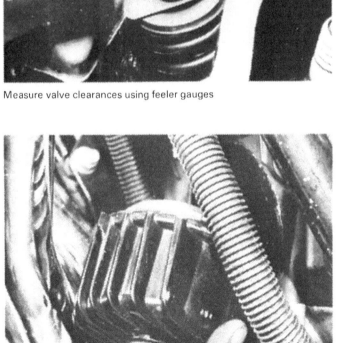

Install oil filter assembly as shown – do not overtighten retaining bolt

Check pad wear line against stop (arrowed)

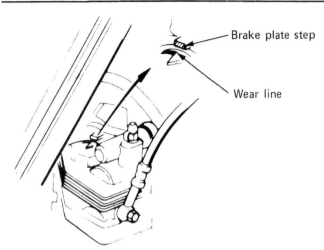

Front brake pad wear check

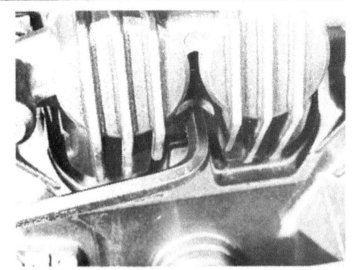

Pads are located by Allen-headed bolt

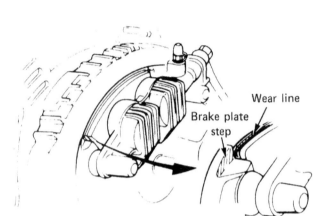

Rear brake pad wear check

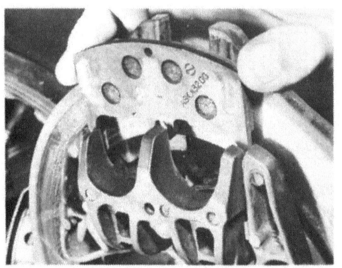

Caliper assembly removed to show arrangement of pads, caliper and disc

The top edge of the pads is visible through the brake plate opening, and each pad has a line denoting the wear limit. If when the brake is applied the line is in line with the step in the brake plate, the pads should be renewed. In the case of the front brake, all four pads should be renewed simultaneously.

The pads can be renewed with the wheel and brake in position. In the case of the front brake, disconnect the speedometer drive cable and the caliper link to the anti-dive unit. Release the three domed nuts which retain the brake shroud and manoeuvre it clear of the brake. Release the pad locating bolt, push the caliper toward the disc to obtain maximum clearance, then lift out the pads.

When fitting new pads it will be necessary to push the caliper pistons inwards to make room. This will cause the fluid level in the reservoir to rise, so check that it does not overflow, and remove surplus fluid if necessary. Once assembly is complete, check the reservoir levels and top up as required.

In the case of the rear brake, check that the brake pedal is set to 10 – 20 mm (0.4 – 0.8 in) below the top of the footrest. If adjustment is required, slacken the master cylinder pushrod adjuster locknut and set the adjuster to give the required height. Check that the rear brake switch operates normally and adjust as required.

Note: At every third service (every 18 months/12000 miles) the hydraulic fluid should be changed as described in Chapter 5. This ensures that the fluid, which degrades in time, is kept to DOT 3/SAE1703J specification and will prolong the life of the brake components.

8 Clutch: adjustment

The clutch lever adjuster should be set to give 10 – 20 mm ($\frac{3}{8}$ – $\frac{3}{4}$ in) free play at the lever end. If there is insufficient movement available, screw the adjuster fully home and reposition the adjuster at the lower end of the cable to take up the excess play.

Cable lower adjuster provides coarse clutch adjustment

9 Checking the electrical system

Check the operation of the lights, turn signals and horn. Renew any blown bulbs (see the appropriate Section of Chapter 6 for details). Check the wiring, looking for signs of chafing or damage. Where necessary, re-route any suspect wiring runs to prevent further wear. If the plastic insulation has worn through, repair it by binding the damaged area with PVC insulating tape. If the wire has been more seriously damaged, cut out the damaged length and solder in a new length, binding the joins with PVC tape. Alternatively, a single bullet type connector can be fitted to join the existing wire where space permits.

Look for signs of water contamination in the various block connectors. This should be dealt with promptly to prevent corrosion of the internal metal parts. A temporary cure can be effected by spraying the connectors with WD40 or a similar silicone-based maintenance spray. A more permanent solution is to dismantle each connector and pack it with silicone grease.

Check that the headlamp aim is set correctly; in most countries traffic laws stipulate the required setting. The alignment operation is described in Section 8 of Chapter 6. Check that the front and rear brake lamp switches operate normally. If necessary, the rear brake lamp switch can be adjusted to come on earlier or later. The front switch is a sealed unit and cannot be adjusted. See Section 17 of Chapter 6 for details.

10 Checking and lubricating the stands

Clean and examine the stand pivots and springs for signs of wear or damage. If necessary, renew worn parts promptly to avoid any risk of failure in use — remember that if the stand drops down when riding it can easily catch in the road surface or a drain cover, throwing the machine out of control. Use only the correct springs and pivot bolts; the latter in particular are specially hardened and shouldered and must not be replaced with an ordinary bolt of similar size. Lubricate the pivots with grease, WD40 or a chain lubricant spray. In the case of the side stand, check that the rubber pad has not worn down to the wear limit line. The pad is designed to flip the stand up if left down accidentally and thus it is essential that it is renewed if worn. The new pad should be marked "OVER 260 lbs ONLY".

11 Checking the suspension and steering

Pump the front suspension up and down, noting any knocking noises and feeling for erratic movement. If play is suspected it must be investigated at once. Place the machine on its centre stand and check for play in the steering head bearings or forks. If any movement is detected establish exactly where the play is located. If there is movement between the fork stanchions and lower legs, or if oil leakage is noted around the fork seals, strip and overhaul the forks as described in Section 4 of Chapter 4.

Check the fork air pressure after the machine has stood overnight. The correct setting is 0.8 – 1.2 kg/cm^2 (11 – 17 psi). If adjustment is required use a hand pump or a syringe type suspension pump — it is easy to over-pressurise the forks when using an air line. Road test the machine and check that the anti-dive system offers progressively greater resistance on each of the four settings. If the system operates erratically or fails to have any effect, strip and overhaul the system as described in Section 5 of Chapter 4.

If play is found at the steering head bearings they can be adjusted provided that the bearings are not worn or damaged. Check this by turning the steering from lock to lock. If there is any sign of roughness or 'notchiness' in the bearings they may have become indented and must be renewed. This is described in Section 3 of Chapter 4. If the bearings are in good condition but in need of adjustment, remove the cap nut at the centre of the top yoke, then adjust the bearing free play using a C-spanner on the slotted adjuster nut just below the top yoke. Tighten the nut until it seats lightly, then back it off by about $\frac{1}{8}$ of a turn. Make any necessary fine adjustments until the steering moves easily from lock to lock but with no discernible free play. Be careful not to overtighten the adjuster. Refit the cap nut and tighten it to 8.0 – 12.0 kgf m (58 – 87 lbf ft).

Check the operation of the rear suspension. If it is suspect in any way, or if there are unusual noises when the suspension is bounced up and down, strip and overhaul the linkage and check the condition of the rear suspension unit. See Sections 7 and 8 of Chapter 4 for details. Although it is not stipulated in the service schedule it is good practice to clean and grease the rear suspension linkage annually to prevent wear developing in the bushes and to draw attention to any developing fault.

12 Checking nuts, bolts and fasteners

Clean the machine thoroughly and inspect all chassis fasteners for security, using the torque wrench settings given in the specification sections of the various Chapters. Check that all split pins and R-pins are in position and undamaged and that all wiring, cable and hose clamps are fitted correctly.

13 Checking the wheels

Clean the wheels thoroughly, paying particular attention to the area around the spoke roots. If corrosion or loose rivets are noted, have the wheel checked by a Honda dealer to eliminate any risk of failure. Check for runout and wheel bearing wear as described in Section 2 of Chapter 5.

Chapter 1 Engine, clutch and gearbox

Contents

Specifications

Engine

Type	Air-cooled four cylinder four-stroke
Bore	59.2 mm (2.33 in)
Stroke	52.0 mm (2.05 in)
Capacity	572.5 cc (35.6 cu in)
Compression ratio	9.5:1

Cylinder identification

Numbered 1, 2, 3, 4 from left-hand side of machine

Cylinder block

Bore diameter	65.500 – 65.520 mm (2.5787 – 2.5795 in)
Service limit	65.60 mm (2.583 in)
Gasket face maximum warpage	0.10 mm (0.004 in)

Pistons and rings

Piston ring to groove clearance:	
Top	0.030 – 0.065 mm (0.0012 – 0.0025 in)
2nd	0.025 – 0.055 mm (0.0010 – 0.0022 in)
Service limit (both)	0.09 mm (0.004 in)
Piston ring end gap:	
Top and 2nd	0.10 – 0.30 mm (0.004 – 0.012 in)
Service limit	0.50 mm (0.020 in)
Oil	0.30 – 0.90 mm (0.012 – 0.035 in)
Service limit	1.1 mm (0.04 in)
Piston diameter (std)	59.170 – 59.190 mm (2.3295 – 2.3303 in)
Service limit	59.10 mm (2.33 in)
Gudgeon pin bore	15.002 – 15.008 mm (0.5906 – 0.5909 in)
Gudgeon pin diameter	14.994 – 15.000 mm (0.5903 – 0.5906 in)
Piston to gudgeon pin clearance service limit	0.04 mm (0.002 in)
Cylinder to piston clearance service limit	0.10 mm (0.004 in)
Piston oversizes	+0.25, 0.50, 0.75, 1.00 mm

Valves

Arrangement	dohc, four valves per cylinder
Valve stem diameter:	
Inlet	4.975 – 4.990 mm (0.1959 – 0.1965 in)

Service limit	4.97 mm (0.196 in)
Exhaust	4.955 – 4.970 mm (0.1951 – 0.1957 in)
Service limit	4.94 mm (0.195 in)
Valve guide internal diameter:	
Inlet and exhaust	5.000 – 5.012 mm (0.1969 – 0.1973 in)
Service limit	5.04 mm (0.198 in)
Valve stem to guide clearance service limit:	
Inlet	0.07 mm (0.003 in)
Exhaust	0.09 mm (0.004 in)
Valve seat width	0.9 – 1.1 mm (0.035 – 0.043 in)
Service limit	1.5 mm (0.06 in)
Valve spring free length:	
Outer	35.24 mm (1.3874 in)
Service limit	34.0 mm (1.34 in)
Inner	31.80 mm (1.2520 in)
Service limit	30.8 mm (1.21 in)

Camshafts and rockers

Cam lobe height service limit:	
Inlet	35.60 mm (1.402 in)
Exhaust	35.44 mm (1.395 in)
Camshaft bearing oil clearance (inlet and exhaust):	
All caps except No 3	0.020 – 0.062 mm (0.0008 – 0.0024 in)
Service limit	0.10 mm (0.0039 in)
Cap No 3	0.050 – 0.092 mm (0.0020 – 0.0036 in)
Service limit	0.15 mm (0.0059 in)
Camshaft runout service limit	0.03 mm (0.0012 in)
Rocker arm bore	12.000 – 12.018 mm (0.4724 – 0.4731 in)
Service limit	12.05 mm (0.4744 in)
Rocker shaft diameter	11.973 – 11.984 mm (0.4714 – 0.4718 in)
Service limit	11.94 mm (0.4701 in)

Cam chain

Service limit	332.00 mm (13.07 in)

Valve timing

	At 1 mm lift	At zero lift
Inlet opens at	10° BTDC	59°20' BTDC
Inlet closes at	35° ABDC	90°14' ABDC
Exhaust opens at	40° BBDC	65°16' BBDC
Exhaust closes at	5° ATDC	65°10' ATDC

Valve clearance

Inlet and exhaust	0.10 – 0.14 mm (0.0039 – 0.0055 in)

Crankshaft and connecting rods

Crankshaft runout service limit	0.05 mm (0.0020 in)
Big-end side clearance	0.05 – 0.20 mm (0.0020 – 0.0079 in)
Service limit	0.3 mm (0.01 in)
Big-end/crankpin clearance	0.020 – 0.055 mm (0.0008 – 0.0022 in)
Service limit	0.07 mm (0.0028 in)
Main bearing/journal clearance	0.020 – 0.045 mm (0.0008 – 0.0018 in)
Service limit	0.06 mm (0.0024 in)
Small-end bearing diameter	15.016 – 15.034 mm (0.5912 – 0.5919 in)
Service limit	15.07 mm (0.5933 in)

Clutch

Free play (at lever end)	10 – 20 mm (0.4 – 0.8 in)
Clutch spring free length	34.2 mm (1.346 in)
Service limit	32.7 mm (1.287 in)
Friction plate thickness	3.22 – 3.38 mm (0.1268 – 0.1331 in)
Service limit	2.90 mm (0.1142 in)
Plain plate warpage service limit	0.3 mm (0.12 in)
Outer bush internal diameter	21.980 – 21.933 mm (0.8654 – 0.8635 in)
Service limit	22.03 mm (0.8673 in)
Outer bush outside diameter	28.015 – 28.028 mm (1.1030 – 1.1035 in)
Service limit	27.97 in (1.1012 in)

Starter clutch

Drive sprocket boss diameter	42.175 – 42.200 mm (1.6604 – 1.6614 in)
Service limit	42.09 mm (1.6571 in)

Gearbox

Type	6-speed constant mesh

Ratios:
  1st .... 2.500 : 1
  2nd .... 1.714 : 1
  3rd .... 1.333 : 1
  4th .... 1.074 : 1
  5th .... 0.931 : 1
  Top .... 0.821 : 1
Gearbox backlash .... 0.046 – 0.140 mm (0.0018 – 0.0055 in)
Service limit .... 0.3 mm (0.012 in)
Gearbox pinion internal diameter:
  Input shaft 5th .... 29.020 – 29.041 mm (1.1425 – 1.1433 in)
  Service limit .... 29.06 mm (1.1441 in)
  Input shaft 6th, Output shaft 1st, 2nd and 3rd .... 28.020 – 28.041 mm (1.1031 – 1.1040 in)
  Service limit .... 28.06 mm (1.1047 in)
Gearbox pinion bush diameter:
  Input shaft 5th ID .... 25.025 – 25.046 mm (0.9852 – 0.9861 in)
  Service limit .... 25.09 mm (0.9878 in)
  Input shaft 5th OD .... 28.979 – 29.000 mm (1.1409 – 1.1417 in)
  Service limit .... 28.94 mm (1.1394 in)
  Input shaft 6th OD .... 27.979 – 28.000 mm (1.1015 – 1.1024 in)
  Service limit .... 27.94 mm (1.1000 in)
  Output shaft 1st ID .... 23.984 – 24.005 mm (0.9443 – 0.9451 in)
  Service limit .... 24.18 mm (0.9520 in)
  Output shaft 1st OD .... 20.000 – 20.021 mm (0.7874 – 0.7882 in)
  Service limit .... 19.97 mm (0.7862 in)
  Output shaft 2nd and 3rd ID .... 25.000 – 25.021 mm (0.9843 – 0.9851 in)
  Service limit .... 25.07 mm (0.9870 in)
  Output shaft 2nd and 3rd OD .... 28.000 – 28.021 mm (1.1024 – 1.1032 in)
  Service limit .... 27.96 mm (1.1008 in)
Input shaft OD at 6th gear .... 24.959 – 24.980 mm (0.9826 – 0.9835 in)
Service limit .... 24.93 mm (0.9815 in)
Output shaft OD at 1st gear .... 19.972 – 19.985 mm (0.7863 – 0.7868 in)
Service limit .... 19.95 mm (0.7854 in)
Output shaft OD at 2nd, 4th gear .... 24.959 – 24.980 mm (0.9826 – 0.9835 in)
Service limit .... 24.93 mm (0.9815 in)
Clearances (service limits):
  Input shaft 5th gear to bush .... 0.12 mm (0.0047 in)
  All others .... 0.10 mm (0.0039 in)
Selector fork thickness .... 5.93 – 6.00 mm (0.233 – 0.236 in)
Service limit .... 5.61 mm (0.221 in)
Selector fork bore diameter .... 12.015 – 12.036 mm (0.473 – 0.474 in)
Service limit .... 12.56 mm (0.494 in)
Selector fork shaft diameter .... 11.966 – 11.984 mm (0.471 – 0.472 in)
Service limit .... 11.90 mm (0.469 in)

Primary drive
  Type .... Hy-Vo chain
  Reduction ratio .... 2.565 : 1

Final drive
  Type .... Sealed endless chain
  Reduction ratio .... 2.812 : 1 (16/45T)

Torque settings

Component	kgf m	lbf ft
Cylinder head cover	0.8 – 1.2	6.0 – 9.0
Camshaft bearing cap	1.0 – 1.4	7.0 – 10.0
Cylinder head:		
6 mm	2.0 – 2.4	14.0 – 17.0
8 mm	2.6 – 3.0	19.0 – 22.0
Camshaft sprocket	1.8 – 2.2	13.0 – 16.0
Spark plugs	1.2 – 1.6	9.0 – 12.0
Crankcase bolts	2.2 – 2.6	16.0 – 19.0
Alternator rotor	4.6 – 5.4	33.0 – 39.0
Connecting rod cap nuts	3.0 – 3.4	22.0 – 25.0
Oil filter bolt	2.8 – 3.2	20.0 – 23.0
Oil pressure switch	1.5 – 2.0	11.0 – 14.0
Neutral switch	1.6 – 2.0	12.0 – 14.0
Sump drain plug	3.5 – 4.0	25.0 – 29.0
Oil feed pipe union bolts	1.0 – 1.4	7.0 – 10.0
Starter clutch retaining bolt	4.8 – 5.2	35.0 – 38.0
Starter clutch bolts	2.6 – 3.0	19.0 – 22.0
Clutch centre nut	4.5 – 5.5	32.0 – 40.0
Engine mounting bolts:		
10 mm	3.5 – 4.5	25.0 – 33.0
8 mm	1.8 – 2.5	13.0 – 20.0

1 General description

The four cylinder engine is built in unit with the primary transmission and gearbox. The light alloy cylinder head features four valves per cylinder operated via forked rocker arms incorporating screw and locknut adjusters from the chain-driven double overhead camshafts. The camshafts run in plain bearing surfaces machined in the cylinder head material and the detachable alloy bearing caps. The cam chain is of the Morse or Hy-Vo type and is controlled by an automatic tensioner.

The horizontally-split crankcase assembly houses a one-piece forged crankshaft supported by five renewable plain bearings. The connecting rods have split big-end eyes, also fitted with renewable bearing inserts. The left-hand end of the crankshaft carries the three-phase alternator in its own compartment, whilst at the right-hand end an elongated chamber contains the ignition pickup assembly and electric starter drive.

Between cylinders 3 and 4 the crankshaft incorporates the primary drive gear, from which power is transmitted to the clutch outer drum via a Hy-Vo chain. The clutch is of the usual wet multi-plate design and incorporates a gear drive for the engine oil pump. The clutch is mounted on the right-hand end of the gearbox input shaft which lies parallel to, and to the rear of, the crankshaft.

The gearbox output shaft is arranged above and to the rear of the input shaft, thus keeping the crankcase assembly as short as possible. The six-speed gearbox is of the constant-mesh type, output being via a sprocket on the left-hand side.

2 Operations with the engine/gearbox unit in the frame

The following components and assemblies may be dismantled with the engine unit in the frame. Where a number of items require attention it may prove preferable to remove the unit to allow easier access on a workbench.

(a) Cylinder head, camshafts and valves*
(b) Starter motor and drive
(c) Ignition pickup and alternator
(d) Gearchange mechanism (external components only)
(e) Clutch assembly
*Refer to the note at the beginning of Section 5 concerning the cylinder head and valves

It will be noted that work on other components or assemblies requires that the unit be removed and the crankcase halves separated. This includes the cylinder block, pistons, crankshaft primary drive, oil pump and gearbox components.

3 Removing the engine/gearbox unit

1 Place the machine securely on its centre stand allowing adequate working space on all sides. It is preferable, though not essential, to raise the machine on a strong bench or horizontal ramp to a comfortable working height.
2 Slide a drain tray beneath the crankcase, remove the sump drain plug and allow the engine oil to drain. The crankcase holds about 3 litre (5.3/6.3 Imp/US pint).
3 Owners of F2 models should remove the fairing at this stage; the engine is awkward to manoeuvre out of the frame and every inch of space is needed. Refer to Chapter 4 for details.
4 Slacken and remove the exhaust pipe retainer nuts. Slide the retainers clear of the cylinder head studs and remove the split collet halves. Release the silencer mounting bolts from the footrest plate. The exhaust system can now be moved forward, then lowered and disengaged from the frame.
5 Unlock and raise the dualseat, then pull off the plastic side panels. Turn the fuel tap off and remove the fuel pipe, vacuum pipe and drain hose. Each is secured by a wire clip which can be slid clear after squeezing together the 'ears'. The pipes can now be worked off with a small screwdriver.
6 Slacken and remove the tank retaining bolt. Lift the rear of the tank, then pull it back to free the front mounting rubbers. Trace and disconnect the fuel gauge sender leads, then remove the tank.
7 Disconnect the ignition coil low tension leads, making a note of the wiring arrangements for future reference. Release the coil bracket mounting bolts and remove the coils together with the spark plug leads and caps.
8 Trace and disconnect the alternator leads at the connector block above the air cleaner casing. Release the regulator/rectifier mounting bolts and remove the unit from the frame. Trace the ignition pickup wiring back to the IC igniter unit, disconnecting it by unplugging the larger of the two connectors.
9 Slacken the single screw which secures the tachometer drive cable and lodge the cable clear of the engine unit. Remove the oil filter bolt and lift away the filter assembly, noting that some oil will escape as the cover is removed. Discard the old filter element. Free the oil cooler hoses at their lower end, then remove the two oil cooler mounting bolts. The assembly can now be removed.
10 Slacken the rear wheel spindle and chain adjusters and push the wheel forward to give maximum chain free play. Remove the gearbox sprocket cover. Slacken and remove the sprocket retaining bolts. Twist the retainer plate until the splines align, then remove it from the shaft. Slide the sprocket and chain clear of the shaft end, then disengage the sprocket, leaving the chain looped around the frame.
11 Disconnect the heavy starter motor lead at the motor terminal. Release the oil pressure switch lead. Slacken the clutch cable adjuster locknuts and free the cable from the support bracket and the release arm. Remove the brake pedal pinch bolt and pull off the pedal.
12 Slacken the large hose clip which retains the two air cleaner casing sections. Slacken the hose clips which retain the carburettors to the air cleaner hoses and to the intake adaptor rubbers. Pull the carburettor assembly rearwards and disengage them from the intake adaptors. The assembly should now be twisted to free the carburettor stubs from the air cleaner hoses, and slid out to the right-hand side. Slacken the choke cable clamp screw and disengage the cable. Slacken the throttle cable adjusters and unhook the cable ends to free them. Remove the carburettor assembly and place it to one side.
13 Disconnect the crankcase breather hose and lodge it clear of the engine unit. Dismantle and remove the rear upper engine mounting plates, freeing the engine earth lead which should be moved clear of the engine. The remaining mounting bolts and plates should now be released, noting that it may prove necessary to take the weight of the engine unit whilst these are withdrawn.
14 The task of lifting the engine unit out of the frame cradle is far from easy due to the very limited clearance available, and requires two people. Lift the unit slightly, then tip the top of the unit over to the right. Check that the front and rear of the crankcase does not foul the frame or brackets, and if necessary manoeuvre the unit around these. The engine is lifted clear of the frame from the right-hand side, angled at about 45° from vertical.

4 Dismantling the engine/gearbox unit: preliminaries

1 Before commencing any dismantling work, the engine unit must be thoroughly cleaned to prevent the ingress of dirt, and also to make the dismantling work cleaner and easier. Plug any open ports or connecting stubs with rag or tape, then work a proprietary degreasing solvent into the engine castings, using a paintbrush. Once all debris has been loosened, hose the unit down and then leave it to dry.
2 Prepare the work area, ensuring adequate bench space, and arrange shelf space to store components in sequence as they are removed. A supply of boxes, tins and plastic bags is invaluable for holding small, easily-lost, components in the correct sequence. A note pad and pen should be to hand so that notes can be made where necessary. A supply of clean, lint-free rags and paper towels should be available.
3 Prior to commencing work, read through the relevant Section to gain some idea of the necessary procedure. At no time should excessive force be required unless specified in the text; if a stubborn fitting or fastener is encountered, re-read the text and study the accompanying illustrations and photographs.

5 Dismantling the engine/gearbox unit: removing the camshafts and cylinder head

Note: Due to poor clearance between the cylinder head and frame attachments, it may be necessary on some machines to either partially or fully remove the engine to enable the cylinder head to be lifted off its studs

1   Slacken and remove the cylinder head cover retaining bolts and remove the cover. If necessary, tap around the joint face to assist in separation.

2   Release the cam chain upper guide bolts and remove the guide. Remove the oil feed pipe union bolts and lift away the internal oil feed pipes.

3   Remove the starter clutch cover to expose the clutch/rotor unit and timing marks. The clutch/rotor assembly can be used to rotate the crankshaft, and thus the camshafts. Turn the crankshaft until one sprocket retaining bolt on each camshaft is accessible, and remove it. Turn the crankshaft through 360° and remove the two remaining bolts. Slide the camshaft sprockets off their shoulders and disengage the sprockets from the chain. **Note:** From this stage onwards do not attempt to rotate the crankshaft.

4   Slacken each of the camshaft cap retaining bolts by about $\frac{1}{2}$ a turn at a time. Valve spring pressure will push the camshafts clear of the head and it is important that the bolts are slackened evenly and progressively to preclude risk of damage. Remove the caps and locating dowels and place them to one side. Disengage the camshafts from the sprockets and chain, then remove the sprockets.

5   Remove the oil pressure switch to free the lower end of the external oil feed pipe, then release the upper end by removing the two union bolts. Release the two external cylinder head bolts, these being located centrally at the front and rear of the head.

6   Slacken evenly and progressively the twelve domed nuts which secure the cylinder head. Pull up the tensioner unit from the cam chain tunnel and release the chain by removing the R-pin and clevis pin which secure the tensioner blade, then place a screwdriver through the chain loop to prevent it from falling into the crankcase. Reassemble the tensioner and place it to one side.

7   The cylinder head can now be lifted off the holding studs. Where necessary, tap around the joint face with a hide mallet to assist in breaking the seal. Note that the cylinder head gasket is of laminated construction and will probably pull apart during removal.

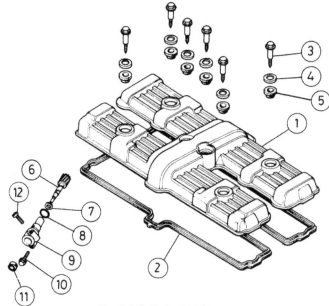

Fig. 1.1 Cylinder head cover

1	Cylinder head cover	7	Washer
2	Cover sealing gasket	8	O-ring
3	Bolt – 6 off	9	Tachometer gear housing
4	Washer – 4 off	10	Bolt
5	Sealing ring – 6 off	11	Oil seal
6	Tachometer gear	12	Screw

Fig. 1.2 Cylinder head

1	Cylinder head	9	End plug – 4 off
2	Cylinder head gasket	10	O-ring – 4 off
3	Spark plug – 4 off	11	External oil feed pipe
4	Exhaust valve guide – 8 off	12	Union bolt – 2 off
5	O-ring – 16 off	13	Sealing washer – 4 off
6	Inlet valve guide – 8 off	14	Bolt
7	Domed nut – 12 off	15	Oil pressure switch
8	Washer – 12 off	16	Union bolt
17	Sealing washer – 2 off	25	Camshaft cap – 4 off
18	Right-hand adaptor – 2 off	26	Locating dowel – 20 off
19	Left-hand adaptor – 22 off	27	Bolt – 16 off
20	O-ring – 4 off	28	Bolt – 4 off
21	Bolt – 8 off	29	Right-hand oil feed pipe
22	Vacuum take off stub	30	Left-hand oil feed pipe
23	Hose clamp – 4 off	31	Union bolt – 4 off
24	Camshaft cap – 6 off		

6   Dismantling the engine/gearbox unit: removing the cylinder block and pistons

1   The cylinder block and pistons can be removed after the camshafts and cylinder head have been detached as described in Section 5. Note that the cylinder block can be removed only after the engine unit has been removed from the frame.
2   Remove the nut and plain washer at the centre of the front edge of the cylinder block. Note that it is possible that a bolt may be found here on some engine units.
3   Lift the cylinder block by an inch or two to allow clean rag to be packed into the crankcase mouths. This will prevent debris falling into the crankcase. If the block is stuck in position, tap around the joint face with a hide mallet to jar it free. Once the rag is in place, lift the block clear of the holding studs, taking care to catch the pistons as they drop free of the bores.
4   Using a small electrical screwdriver or a pair of pointed-nose pliers, displace and remove the gudgeon pin circlips on one side of each piston. The piston can now be freed by displacing the gudgeon pin. As each piston is removed, mark it, after cleaning, with a spirit-based felt marker to indicate the bore from which it was removed. Finally, remove the two locating dowels and place them with the cylinder block for safe keeping.

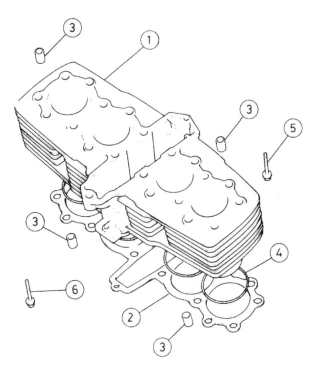

Fig. 1.3 Cylinder block and pistons

1	Cylinder block	6	Bolt
2	Cylinder base gasket	7	Piston
3	Locating dowel – 4 off	8	Piston rings
4	O-ring – 4 off	9	Gudgeon pin
5	Bolt	10	Circlip – 2 off

7   Dismantling the engine/gearbox unit: removing the alternator

1   Remove the engine left-hand outer cover by releasing its three retaining bolts. The alternator stator assembly is mounted inside the cover and need not be disturbed unless specific attention is required. If the engine unit is in the frame and it is wished to remove the stator completely, trace back and separate the wiring connector.
2   Slacken and remove the rotor retaining bolt. If it proves tight, hold the rotor with a strap wrench to prevent crankshaft rotation. A large thread is provided to allow an extractor bolt to be used to draw off the rotor. This tool, Part Number 07933-4250000, is available from Honda dealers. Alternatively, a legged puller may be used, but note that it **must** bear only on the inner face of the rotor or damage will result.
3   When the tool is in place, tighten down the centre bolt firmly and then strike the bolt head sharply with a hammer. This should jar the rotor free.

7.2 Legged puller may be used to draw off the rotor

8   Dismantling the engine/gearbox unit: removing the starter drive, starter motor and ignition pickup assembly

1   Release the eight starter drive cover retaining bolts and lift away the cover. Prevent crankshaft rotation by passing a screwdriver blade between two of the starter clutch bolt heads, then slacken and remove the central retaining bolt. The starter clutch, drive chain and drive sprocket can now be withdrawn as an assembly. Remove the pulser coil retaining bolts and remove the coils together with the chain guide. Note that the wiring protector plate should be released to free the wiring between the two pulser coils.
2   Remove the two bolts which secure the starter motor to the crankcase, then pull the motor clear. The short shaft which transfers drive from the motor spindle to the sprocket can be withdrawn from its housing.

9   Dismantling the engine/gearbox unit: removing the clutch

1   If this operation is being undertaken with the engine unit in the frame, note that it will first be necessary to remove the clutch cable and brake pedal, and to drain the engine oil.
2   Remove the bolts which secure the clutch outer cover and lift the cover away together with its gasket. Slacken and remove the four bolts which secure the clutch release plate. Remove the plate and bearing but leave the springs in position. Refit and tighten the four bolts, having first placed plain washers beneath their heads.
3   The clutch centre nut can now be slackened, noting that it will be necessary to prevent the clutch assembly from rotating. If the engine unit is in the frame, select 1st gear and refit the brake pedal which can

then be applied whilst the nut is removed. If the engine is being stripped on a workbench, lock the crankshaft by passing a smooth round metal bar through one of the connecting rod small-end eyes, resting the ends on wooden blocks placed across the crankcase mouth. The clutch centre, pressure plate and the plain and friction plates can now be removed as a unit. Note that the clutch outer drum cannot be removed until the crankcase halves have been separated.

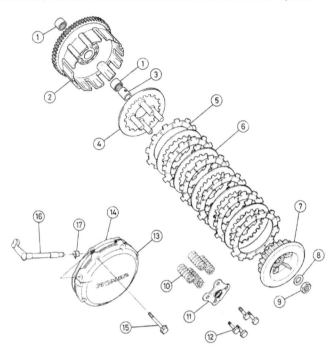

Fig 1.4 Clutch

1	Clutch centre needle roller bearing – 2 off	9	Centre nut
2	Outer drum	10	Spring – 4 off
3	Sleeve	11	Release plate
4	Pressure plate	12	Bolt – 4 off
5	Friction plate – 6 off	13	Clutch outer cover
6	Plain plate – 5 off	14	Cover gasket
7	Clutch centre	15	Bolt – 8 off
8	Washer	16	Operating lever
		17	Spring

10 Dismantling the engine/gearbox unit: removing the external gearchange components

1   Remove the gearchange pedal and disconnect the lead from the neutral switch. Remove the cover retaining bolts, noting that where this operation is undertaken with the engine unit in the frame, it will first be necessary to drain the engine oil. As the cover is lifted away, remove the locating dowels and place them with the cover.

2   Release the cam plate retaining bolts and remove the cam plate components from the end of the selector drum, leaving the stopper plate in position. Remove the stopper arm pivot bolt and lift away the arm and spring. Grasp the end of the selector claw assembly. The stopper plate can now be removed.

11 Dismantaling the engine/gearbox unit: separating the crankcase halves

1   Slacken and remove the upper crankcase bolts, then invert the engine unit on the workbench. Remove the sump retaining bolts, slackening them evenly and progressively to avoid warpage. Lift away the sump. Pull off the oil strainer, the bypass pipe and the relief valve, these being a push fit and sealed by O–rings.

2   Slacken the 6 mm crankcase bolts in a criss–cross sequence by about ½ a turn at a time until all are loose, then slacken the ten 8 mm

main bearing bolts in a reverse of the tightening sequence in Fig 1.12 later in this Chapter; again slacken by ½ turn at a time. The lower crankcase half can now be lifted clear together with the selector drum and forks and the oil pump. If the crankcase halves prove reluctant to separate, try tapping around the joint face with a hide mallet. On no account attempt to lever between the cases to assist separation because the sealing surfaces will be damaged.

3   Lift out the primary chain oil nozzle and place it in a bag or box to avoid its loss. Lift the clutch drum and gearbox input shaft by an inch or so and slide the shaft clear. The clutch drum can now be disengaged from the primary chain and removed, as can the crankshaft complete with primary and cam chains.

4   To remove the gearbox output shaft it will first be necessary to detach the bearing support block. This is retained by a pin which in turn is held in place by a bolt, located adjacent to the arrow mark on the support block. Remove the bolt and screw a 5 mm bolt into the threaded end of the pin. Grasp the bolt with pliers and pull it and the pin clear. Using a screwdriver in the prising point in the bearing support block, displace the latter until it is clear of the crankcase.

5   Slide the 1st and 5th gear pinions off the output shaft. Remove the circlip which secures the 3rd gear pinion and remove it. Twist and release the locking washer arrangement to free the 2nd gear pinion. The output shaft can now be slid clear of the bearing support block hole and removed. The crankcase breather oil separater cover and the primary chain tensioner need not be removed unless specific attention is required.

6   Moving to the upper casing half, release the bolts and retainers which locate the selector fork shafts, then withdraw each shaft and remove the forks. Displace and remove the selector drum. Pass a screwdriver through one of the holes in the oil pump pinion and lock it against the casing. Remove the pinion retaining bolt and the pinion. From the underside of the casing, release the three oil pump mounting bolts. The pump can now be manoeuvred clear of the crankcase.

11.4   5 mm bolt or screw can be used to pull out locating pin

12 Examination and renovation: general

1   Before examining the parts of the dismantled engine unit for wear it is essential that they should be cleaned thoroughly. Use a petrol/paraffin mix or a high flash-point solvent to remove all traces of old oil and sludge which may have accumulated within the engine. Where petrol is included in the cleaning agent, normal fire precautions should be taken and cleaning be carried out in a well ventilated place.

2   Examine the crankcase castings for cracks or other signs of damage. If a crack is discovered it will required a specialist repair.

3   Examine carfully each part to determine the extent of wear, checking with the tolerance figures listed in the specifications section of this chapter or in the main text. If there is any doubt about the condition of a particular component, play safe and renew.

4   Use a clean lint free rag for cleaning and drying the various

components. This will obviate the risk of small particles obstructing the internal oilways and causing the lubrication system to fail.

5   Various instruments for measuring wear are required, including a vernier gauge or external micrometer and a set of standard feeler gauges. The machine's manufacturer recommends the use of Plastigage for measuring radial clearance between working surfaces such as shell bearings and their journals. Plastigage consists of a fine strand of plastic material manufactured to an accurate diameter. A short length of Plastigage is placed between the two surfaces, the clearance of which is to be measured. The surfaces are assembled in their normal working positions and the securing nuts or bolts fastened to the correct torque loading; the surfaces are then separated. The amount of compression to which the gauge material is subjected and the resultant spreading indicates the clearance. This is measured directly, across the width of Plastigage, using a pre-marked indicator supplied with the Plastigage kit. If Plastigage is not available, both an internal and external micrometer will be required to check wear limits. Additionally, although not absolutely necessary, a dial gauge and mounting bracket is invaluable for accurate measurement of end float, and play between components of very low diameter bores – where a micrometer cannot reach.

6   After some experience has been gained, the state of wear of many components can be determined visually or by feel and thus a decision on their suitability for continued service can be made without resorting to direct measurement.

13 Examination and renovation: crankshaft, big-end and main bearings and connecting rods

1   Measure the side clearance between each connecting rod big-end and its adjacent crankshaft thrust face, using feeler gauges. If the clearance exceeds the service limit of 0.03 mm (0.01 in) the connecting rod assemblies will require renewal. Clean and mark each connecting rod with a spirit-based felt marker to indicate the cylinder to which it belongs, then remove the big-end cap nuts and release the rods from the crankshaft. The rods and caps are matched and should not be interchanged.

2   Examine the big-end and main bearing journals for signs of scoring or scuffing. Any damage of this nature indicates the need for crankshaft renewal. Note that undersized bearing shells (inserts) are not available and this effectively rules out re-grinding. If it is felt that crankshaft renewal is necessary it is advisable to have this diagnosis confirmed by a Honda dealer.

3   It is good practice to renew the big-end and main bearing inserts during an overhaul. If, however, it is wished to check the bearing clearances this can be done using Plastigage as described in Section 12, or by direct measurement. The clearance service limits are 0.07 mm (0.0025 in) for the big-end bearings and 0.06 mm (0.0024 in) in the case of the main bearings.

4   Crankshaft runout is checked by placing the crankshaft ends on V-blocks and measuring the runout at the centre journal using a dial gauge, noting that $\frac{1}{2}$ of the total reading shown represents the actual runout figure. This must not exceed the 0.05 mm (0.002 in) service limit. If the crankshaft proves serviceable, refer to Section 14 for details on bearing selection.

5   Examine the connecting rods for indications of cracking. This is unlikely to be discovered unless severe engine damage has been sustained due to seizure or a dropped valve, but where it is apparent the affected rod or rods must be renewed, together with the crankshaft if this, too, is bent. If a bore micrometer is available, check the small-end eye internal diameter. If this exceeds the service limit of 15.07 mm (0.593 in) the rod must be renewed.

14 Examination and renovation: big-end and main bearing insert selection and connecting rod installation

1   The criteria for bearing insert selection are various marked dimensions on the crankshaft, connecting rods and crankcase:

(a) *Crankshaft big-end journal OD code. This takes the form of a code **letter** stamped on the adjacent crankshaft web.*
(b) *Connecting rod big-end ID code. This consists of a code **number** stamped on the edge of each connecting rod and cap.*

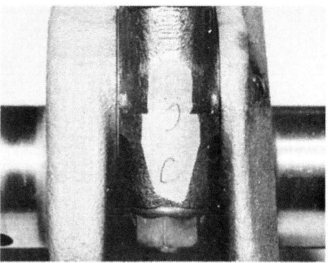
13.1a Take note of big end match marks during removal

13.1b Connecting rods may be removed for examination

(c) *Crankshaft main bearing journal OD code. This is a code **number** just below the big-end OD code letter.*
(d) *Crankcase main bearing journal ID code. These are marked as* **I, II, III or X** *above the left-hand main bearing boss.*

2   To select the appropriate big-end insert, check the crankshaft big-end OD letter and the connecting rod big-end ID number, then use the table below to select the colour-coded insert.

Connecting rod ID code		Crankshaft big-end OD code		
Code	Size	**A** 31.992-32.000mm	**B** 31.984-31.992mm	**C** 31.976-31.984mm
1	35.000-35.008mm	Yellow	Green	Brown
2	35.008-35.016mm	Green	Brown	Black
3	35.016-35.024mm	Brown	Black	Blue

3   Main bearing inserts are selected on a similar basis to that described above, using the table shown below.

42

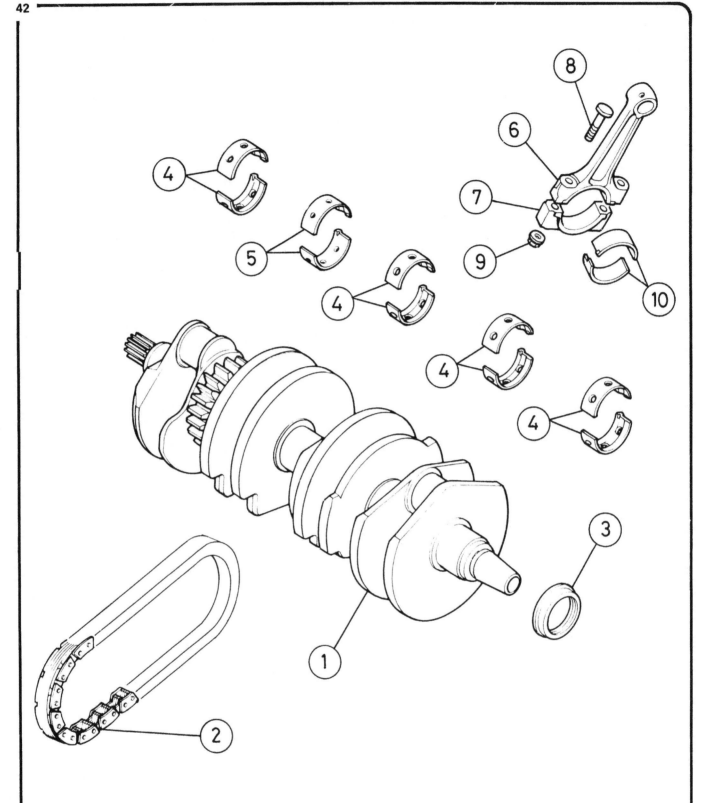

Fig. 1.5 Crankshaft

1 Crankshaft
2 Primary chain
3 Oil seal
4 Main bearing shells
   – 8 off
5 Main bearing shells
   – 2 off
6 Connecting rod – 4 off
7 Big-end cap – 4 off
8 Bolt – 8 off
9 Nut – 8 off
10 Big-end bearing
   shells – 8 off

Crankcase ID   Crankshaft main bearing journal OD code
code

Code \ Size	1 31.994- 32.000mm	2 31.988- 31.994mm	3 31.982- 31.988mm	4 31.976- 31.982mm
I 35.000- 35.006mm	red	pink	yellow	green
II 35.006- 35.012mm	pink	yellow	green	brown
III 35.012- 35.018mm	yellow	green	brown	black
X 35.018- 35.024mm	green	brown	black	blue

4   If in any doubt about calculating the correct insert colours when ordering new parts, make a note of the marked dimensions described in paragraph 1 and give these to your Honda dealer when placing the order.

5   Fit the new big-end bearing inserts to the connecting rods and caps, ensuring that the colour codes are correct. Check that the locating tabs engage correctly and that the oil feed holes are aligned with those in the connecting rods. Coat the bearing faces with molybdenum disulphide grease, and assemble each rod on its crankpin, making sure that the oil hole faces forward. Tighten the cap nuts evenly in two or three stages to 3.0 – 3.4 kgf m (22 – 25 lbf ft), then check that the connecting rods move smoothly round the crankpins.

15   Examination and renovation: checking the cam chain

1   Place the cam chain around the two camshaft sprockets as shown in the accompanying illustration. Using a spring balance or weights apply a pull of 13 kg (28.7 lb) on the chain, then measure across the distance indicated in the illustration. Renew the chain if the measured distance exceeds 332 mm (13.07 in). If there is any doubt about the condition of the cam chain compare its length with that of a new chain.

16   Examination and renovation: cylinder bores and pistons

1   Examine the cylinder bore surfaces for scoring or scratching. Scuffing and 'picking up' of the bores is often the result of partial or complete seizure. If the bores are obviously damaged it will be necessary to have the block rebored and honed to suit the next oversize of piston, and further examination will not be necessary.

2   Measure bore wear using a bore micrometer – most Honda dealers will do this for you – at the top, centre and bottom of the bore. Take each measurement along the gudgeon pin axis and at 90° to this. On a standard bore size of 65.000 – 65.020 mm (2.5787 – 2.5795 in) the service limit is 65.60 mm (2.583 in). A similar wear allowance of about 0.60 mm (0.024 in) can be applied to subsequent oversizes. A rebore will be required if one or more bores have reached or exceeded the service limit.

14.1 Main bearing code is stamped on crankcase as shown

14.5 Tighten cap nuts evenly to prescribed torque setting

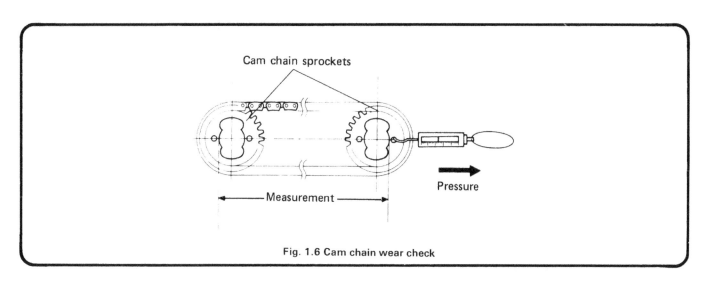

Fig. 1.6 Cam chain wear check

3   If a rebore is deemed necessary, oversize pistons are available in increments of 0.25 mm (0.010 in) up to +1.00 mm (+0.040 in). It is vital that the reboring is carried out by a competent engineering company, and to this end it is recommended that the work should be entrusted to a Honda dealer who will have the necessary contacts.

4   If the bores are in good condition, check that pistons are serviceable. Start by checking the clearance between the top and second piston rings and their respective grooves using feeler gauges. If the clearance exceeds the service limit of 0.09 mm (0.004 in), new rings will normally resolve the problem, though where very high mileages have been covered, the ring grooves may have enlarged, necessitating piston renewal. Note that it is inadvisable to fit new rings to a part-worn bore because of the risk of breakage against the wear ridge at the top of each bore.

5   Remove the rings from each piston, marking the rings to denote the piston to which they belong and the top face of each one. The rings are brittle and easily broken, and are best removed by the method shown in the accompanying illustration. Insert each ring in turn into its correct bore, using the piston to position it about ½ inch up from the bottom. Measure the end gap of the rings, comparing the readings obtained with those shown in the Specifications. Renew the rings as a set if worn to or beyond the service limit.

6   Examine the piston skirt area for signs of wear or damage. Scoring is unlikely unless oil changes have been neglected, whilst scorch marks below the piston ring grooves indicate blow-by of gases from the combustion chamber, often due to bore or piston ring wear. If the engine has suffered seizure, the thrust faces of the piston may have 'picked-up', smearing piston material down the skirt faces. Such damage, if minor, may be smoothed off with fine abrasive paper.

7   Measure the outside diameter of the piston near the bottom of the skirt and at 90° to the gudgeon pin axis. The service limit is 59.10 mm (2.33 in). Subtract this measurement from the measured bore diameter, described in paragraph 2 above, to obtain the piston to bore clearance. Reboring and new pistons will probably be required if this exceeds the service limit of 0.1 mm (0.004 in). If in doubt have the measurement double-checked by a competent motorcycle service agent.

8   Measure the gudgeon pin diameter, and the piston gudgeon pin bore diameter, comparing the readings with those quoted in the Specifications. Calculate from the above measurements the gudgeon pin to piston clearance. If this exceeds the specified service limit, or if the pin is obviously loose in the piston, the piston assembly must be renewed.

9   If new pistons or rings are to be fitted to a part-worn bore, note that it is important that the wear ridge near the top of each bore is removed to avoid ring breakage. It is also necessary to roughen the glazed bore surfaces to allow the new rings to bed in. If necessary, have this work carried out by a Honda dealer.

17 Examination and renovation: camshafts and rocker arms

1   Examine the camshaft bearing surfaces for signs of scoring or other damage. Check the cam lobes for wear, noting that if the hardened surface has worn through or picked up it will be necessary to renew the camshaft.

2   Support the camshaft ends on V-blocks and check for run-out using a dial gauge. If this exceeds the 0.05 mm (0.002 in) service limit the camshaft must be renewed.

3   The camshaft runs directly in the cylinder head material, therefore any damage or excessive wear of the bearing surfaces will require the renewal of the camshaft(s) and possibly the cylinder head as well. The camshaft to cylinder head clearance can be checked by direct measurement or by using Plastigage. If in any doubt about the condition of the bearing surfaces, it is recommended that a Honda dealer is consulted.

4   The rocker arms pivot on two shafts running parallel to the camshafts. Each shaft is located by tapered pins and the ends of the shaft bores are closed by Allen-headed cap bolts. The locating pins can be drawn out using the appropriate service tool, Part Number 07936-MA70000. Alternatively, lock a pair of self-grip pliers over the pins and lever the pin free using a screwdriver, taking care not to mark the gasket face.

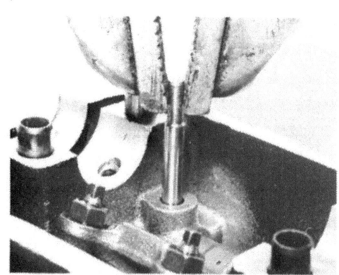

17.4 Grasp locating pins with self-grip pliers then lever out

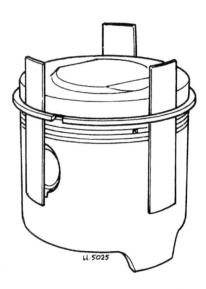

Fig 1.7 Method of removing piston rings

17.5a Remove rocker shaft cap bolts ...

17.5b ... and use screw to help extract shaft

17.5c The shaft can now be pulled clear of the head ...

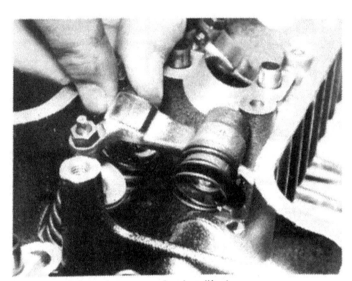

17.5d ... and the rocker arms and springs lifted away

5   Remove the cap bolts and screw a 6 mm bolt into the end of the pivot shaft, using the bolt to draw the shaft clear of the cylinder head. Remove each rocker and spring in turn, placing them in a marked container.

6   Inspect the rocker arm bores for signs of wear, paying particular attention to the hardened pads which bear on the camshaft lobes. Measure the rocker arm bores and the corresponding shaft diameter. If these exceed the appropriate service limits they must be renewed.

Fig. 1.8 Camshafts and rocker arms

1 Exhaust camshaft	11 Spring retainer – 16 off
2 Inlet camshaft	12 Outer valve spring
3 Right-hand rocker arm	– 16 off
shaft – 2 off	13 Inner valve spring
4 Left-hand rocker arm	– 16 off
shaft – 2 off	14 Outer spring seat
5 Locating pin – 4 off	– 16 off
6 Rocker arm – 8 off	15 Inner spring seat
7 Spring – 8 off	– 16 off
8 Locknut – 16 off	16 Valve stem seal – 16 off
9 Adjusting screw	17 Exhaust valve – 8 off
– 16 off	18 Inlet valve – 8 off
10 Collet halves – 16 off	

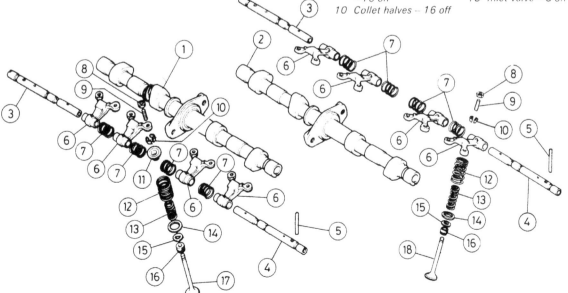

18 Examination and renovation: cylinder head and valves

1   Remove the rocker arms and shafts as described in Section 17 to gain access to the valves. Using a valve spring compressor, compress each valve spring assembly in turn. Displace and remove the collet halves, release the compressor and remove the retainer and valve springs. The valve can now be displaced and removed. Repeat this sequence to remove the remaining valves, placing each one with its springs, retainer and collet halves in marked containers.

2   Examine the valve seating faces for pitting or cracking. If severe, the valve in question should be renewed, whilst light pitting or discolouration can be removed by grinding. Check the corresponding seating area of the valve seats. Light pitting can again be dealt with by grinding, but deeper indentations will necessitate recutting the valve seats. If this proves necessary, check, and where appropriate renew, the valve guides **before** recutting.

3   Examine the valve stems, rejecting any valve which is obviously bent or has deep score marks along the stem surface. With the valve in position, feel for free play between it and the guide. If movement is excessive it will be necessary to renew the guide. If measuring equipment is available, check the valve stem OD and the valve guide ID, subtracting the former from the latter to give the stem-to-guide clearance. If this exceeds the service limit, renew the guide and valve as appropriate.

4   If the above checks show that valve guide renewal is required, it is suggested that the job is entrusted to a Honda dealer who will have the tools and facilities to renew the guides, ream them to size and to re-cut the valve seats to suit.

5   If the valves are in serviceable condition, the contact faces of each valve and seat can be re-faced by grinding. Use a rubber sucker type grinding tool to lap the valve against the seat in an oscillating motion, using fine carborundum grinding paste. Lift the valve frequently to redistribute the paste. When grinding is complete, the valve and seat should each show an even and unbroken matt grey band around the contact area. Be sure to remove all traces of grinding paste when the job is finished. Clean any loose carbon from the ports and then wash thoroughly with paraffin or a degreasing solvent.

6   Examine all valve springs, rejecting any showing obvious signs of damage or fatigue. Measure the spring free lengths and renew any that fall below the service limit.

7   When fitting the valves, check that the valve stem oil seals are in good condition. It is advisable to renew these as a precaution during an overhaul. Lubricate the seal lips and valve stems with molybdenum disulphide grease prior to installation, and take care not to damage the seal as the valve is inserted. Fit the valve springs with the tighter wound coils towards the cylinder head. Once the collet halves are in place, remove the compressor and tap the valve stem lightly to ensure that they are seated fully.

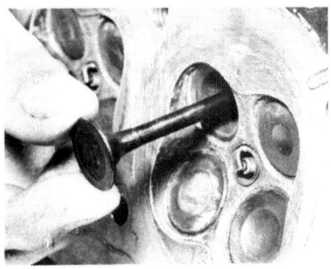
18.1a Compress valve springs to release collet halves

18.1b Retainer and valve springs can be lifted away ...

18.1c ... and valve displaced for examination

18.6a Do not omit valve spring seats during assembly

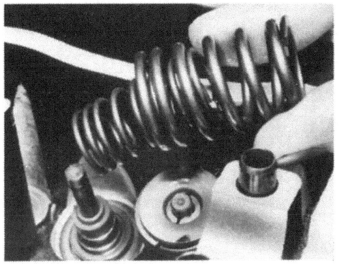

18.6b Fit valve springs with tighter coils downward

19 Examination and renovation: starter drive and clutch

1 The starter drive comprises a chain and two sprockets, one of which forms the inner sleeve of the starter clutch unit. Wear of the drive components is entirely dependent on the frequency with which the starter is used, but is unlikely to prove to be a significant problem. No wear limits are prescribed, but if the sprockets appear worn or hooked or if the chain has stretched when compared with a new one, renew the affected parts.

2 Twist and remove the driven sprocket from the clutch unit. Examine the outside diameter of the boss for wear or indentation. If the surface of the boss appears smooth and even, check the outside diameter. Renewal will be necessary if it exceeds the service limit of 42.09 mm (1.657 in). Check that the rollers are smooth and free from flats, and that they move smoothly against spring pressure.

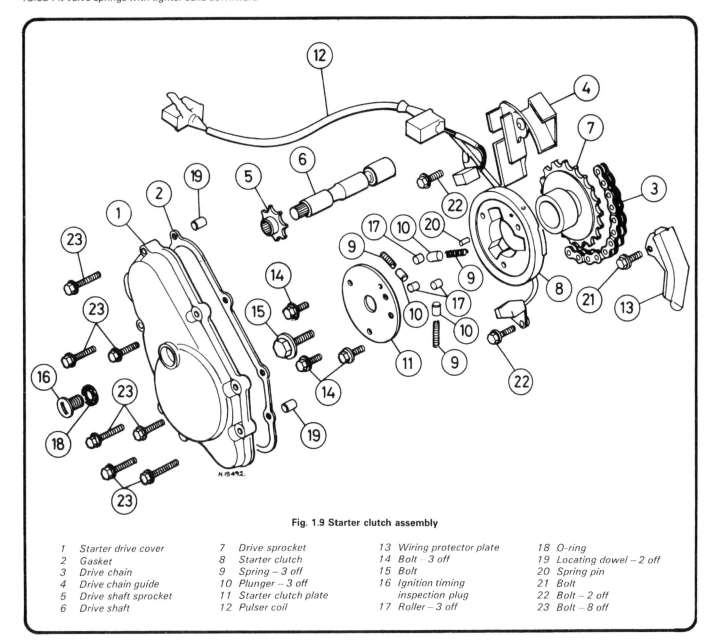

Fig. 1.9 Starter clutch assembly

1	Starter drive cover	7	Drive sprocket	13	Wiring protector plate
2	Gasket	8	Starter clutch	14	Bolt – 3 off
3	Drive chain	9	Spring – 3 off	15	Bolt
4	Drive chain guide	10	Plunger – 3 off	16	Ignition timing
5	Drive shaft sprocket	11	Starter clutch plate		inspection plug
6	Drive shaft	12	Pulser coil	17	Roller – 3 off

18	O-ring
19	Locating dowel – 2 off
20	Spring pin
21	Bolt
22	Bolt – 2 off
23	Bolt – 8 off

20 Examination and renovation: clutch assembly

1   Measure the overall free length of the clutch springs using a vernier caliper. If one or more springs has compressed to the service limit of 32.7 mm (1.28 in) or less, renew the springs as a set.
2   Measure the friction plate thicknesses using a vernier caliper. Renew any plate which has worn to the service limit of 2.90 mm (0.11 in) or beyond. Check the plain plates for warpage on a surface plate. The service limit is 0.3 mm (0.012 in).
3   Examine the slots in the clutch drum and the clutch centre for indentation or burring. Small amounts of damage can be removed by judicious filing; more severe damage will require the renewal of the affected component.
4   Inspect the clutch centre bearing for excessive play or damage. It is not possible to assess wear of the bearing by direct measurement, but any marking or flattening of the rollers is indicative of the need for renewal. A worn bearing can be removed by driving it out using a socket of a diameter fractionally less than that of the bearing outer. Fit the new bearing in a similar fashion, taking care not to distort the outer race.

5   The sleeve on which the bearing runs should be free from scoring or indentation. Measure the sleeve internal diameter and external diameter, renewing it if it is beyond the service limits of 22.03 mm (0.867 in) and 27.97 mm (1.012 in) respectively.
6   The cam-type clutch release mechanism is unlikely to require attention, but may be removed from the cover after freeing the circlip and spring which locate its lower end. The mechanism should be greased prior to assembly.

21 Examination and renovation: gearbox components

1   Dismantle the gearbox input and output shafts separately to avoid any confusion during reassembly. The shafts can be stripped by following in reverse order the assembly sequence of photographs which accompany this section. Note that the output shaft will have been partly stripped during removal. Lay each component out on a clean surface in the order in which it was removed.
2   Inspect the teeth of each gearbox pinion in turn. In normal use, the load faces of each tooth will become lightly polished in appearance. If pitting of the hardened faces is evident, the affected pair of gears should be renewed. The engagement dogs and their corresponding slots or holes should be clearly defined. Any clipping or rounding off will affect engagement and will require renewal of the gear.
3   Referring to the Specifications at the beginning of the Chapter, measure the various internal and external diameters listed, renewing any component which is beyond its service limit. If wear is suspected, but the necessary measuring equipment is not available, have the gearbox components checked by a Honda dealer. If in doubt as to the condition of any of the gearbox shaft components, play safe and renew it; this may save further expense and considerable extra work at a later date.
4   Check the oil seal and O-ring in the output shaft support block, renewing both if marked or damaged in any way. Note that it is good practice to renew them as a precaution during an overhaul.
5   Examine the selector drum, forks and shafts for signs of wear or damage, paying particular attention to the selector fork grooves in the drum. Serious wear is unlikely unless a very high mileage has been recorded. Measure the selector fork claw thickness and internal diameter, renewing them if they exceed their service limits.
6   Assemble the gear shafts, following the accompanying photographic sequence. Lubricate the thrust faces of each gear with molybdenum disulphide grease during assembly, and ensure that all oil holes are aligned as shown in the photographs.

20.3 Bearing needle rollers should be free from indentations or discolouration

21.6a Fit headed bush over input shaft as shown

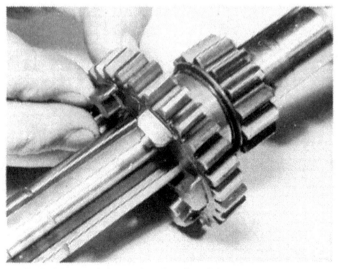

21.6b Fit 5th gear pinion, noting direction of dogs

21.6c Slide splined thrust washer over shaft and secure

21.6d Combined 2nd/3rd gear pinion is fitted as shown

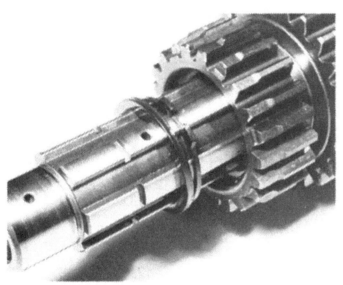

21.6e Fit circlip followed by splined thrust washer

21.6f Oil hole in splined sleeve must align with hole in shaft

21.6g 6th gear pinion can now be positioned

21.6h Place splined thrust washer over shaft ...

21.6i ... and turn as shown to secure pinion

21.6j Special locking washer will now secure thrust washer

21.6k Fit thick plain thrust washer and 4th gear pinion

21.6l Lubricate and fit needle roller bearing ...

21.6m ... followed by outer race

21.6n Press bearing over r.h. end of shaft ...

21.6o ... then fit thrust washer and sleeve

the short oil feed pipe in its recess between the primary drive sprockets, noting the locating pin. Place it in its recess, next to the input shaft left-hand bearing, the oil restrictor orifice. Fit a new O-ring to the adjacent oil feed stub.

4   Moving to the crankcase lower half, fit the main bearings in their recesses and lubricate with engine oil. Slide the selector drum into position in the casing, then tip the casing onto its side to allow access to the selector mechanism recess. Place the detent cam on the end of the selector drum, then fit the bearing retainer and the detent stopper arm. Place the selector claw assembly in its recess, noting that the two ends of the centralising spring must locate on either side of the locating pin. Fit the selector pins and end plate and secure the central retaining bolt. Check the operation of the selector mechanism, then select the neutral gear position, noting that the neutral switch contact blade should be near the 12 o'clock position.

5   Manoeuvre the oil pump into position through the sump aperture and retain it with its three holding bolts. Fit the selector forks and shafts as shown in the accompanying photograph, then retain them with the bolt and retainer plate at the end of each shaft. The pump drive pinion can now be fitted and its retaining bolt secured. Check, and where necessary refit, the primary chain guide assembly and the cam chain tensioner blade socket.

22 Engine reassembly: general preparation

1   Before commencing reassembly, check that each component has been cleaned and that every gasket face is free from old gaskets or jointing compound. Removal of the latter is greatly facilitated by the use of a solvent. This varies according to the composition of the compound, but methylated spirit, acetone or cellulose thinners are often effective. In some cases scraping may be the only answer, but take great care not to damage the gasket face itself.

2   Check that all gaskets, seals and O-rings are to hand and that they are of the correct type for the model in question; a dry run to check this point may save much time and frustration later. Clear the workbench of all unnecessary tools or parts, and make ready an oil can filled with clean engine oil for lubrication as assembly proceeds.

3   Make sure that a torque wrench is available and refer to the torque wrench settings at the front of this Chapter. Note that failure to observe torque settings may result in oil leakage, broken fasteners or distorted castings.

23 Engine reassembly: refitting the crankcase components

1   With the 4th and 6th gear pinions in place, slide the output shaft partly into position through the bearing housing hole, noting that the oil hole in the 6th gear pinion should align with the corresponding hole in the shaft. Fit the 2nd gear pinion. Fit the special thrust washer, then retain this by turning it slightly to allow the lock washer to be installed (see photograph). Fit the 3rd gear bush and pinion, ensuring that the oil holes align, and retain it with its circlip. Fit the 5th gear pinion, again aligning the oil holes, then fit the 1st gear pinion and bush, followed by its thrust washer. The shaft assembly can now be slid fully into position. Fit the bearing retainer block, ensuring that the arrow mark on its edge is aligned with the retaining pin hole. Drop the pin into place and retain it by tightening the retaining pin bolt.

2   Fit the main bearing inserts into their recesses, ensuring that the locating tabs engage properly. Fit the primary drive and camshaft chains around the crankshaft. Lubricate and fit the oil seal to the left-hand (alternator) end of the crankshaft, then lower the assembly into position. Make sure that the seal and bearing inserts locate properly.

3   Lubricate the clutch drum needle roller bearings, then loop the primary chain around the clutch drum and place it in its recess in the crankcase. Fit the clutch bearing sleeve over the end of the assembled input shaft, lubricating it thoroughly with engine oil. The shaft can now be slid through the clutch and lowered into the casing recesses. Ensure that the input shaft bearings locate fully over the pin and half-ring. Fit

23.1a The output shaft, with bearing and 4th gear pinion

23.1b Slide 6th gear pinion over shaft, selector groove as shown

23.1c Fit thrust washer and 2nd gear bush

23.1d Fit shaft through casing and install 2nd gear pinion

23.1e Position special thrust washer, followed by locking washer

23.1f 3rd gear pinion runs on headed bush ...

23.1g ... the oil holes of which must align with those of shaft

23.1h Retain with circlip as shown

23.1i Fit 5th gear pinion, aligning the oil holes, then fit 1st gear bush ...

23.1j ... followed by 1st gear pinion and thrust washer

23.1k Check that seal is in good condition ...

23.1l ... before installing bearing retainer block

23.1m Check that arrow marks align, then fit retaining pin

23.2a Refit oil separator plate using Loctite on bolts

23.2b Fit the primary chain tensioner (2 bolts) ...

23.2c ... and place tensioner blade in position shown

23.2d Lubricate and fit bearing inserts ...

23.2e ... fit cam and primary chain, then install crankshaft

23.3a Check that locating dowel is in position ...

23.3b ... and fit the half-ring in its groove

23.3c Place the clutch drum in its casing recess

23.3d Install the gearbox input shaft

23.3e Check gear operation and grease selector grooves

23.3f Check that locating pin is in place ...

23.3g ... and fit the oil feed nozzle

23.3h Fit a new O-ring to oil feed stub

23.3i ... and fit oil restrictor orifice

23.4a Slide the selector drum into its casing recess

23.4b Fit the detent cam, noting location pin and slot

23.4c Place retainer plate in position and fit single bolt ...

23.4d ... then fit detent stopper arm and pivot bolt

23.4e Fit selector pins and secure end plate and neutral contact

23.4f Note that centralising spring end must locate as shown

23.5a Fit oil pump locating dowels ...

23.5b ... then manoeuvre the pump into casing recess and secure

23.5c Position oil pump drive pinion and secure with central bolt

23.5d Refit the selector forks and shafts ...

23.5e ... and secure the shaft retainers

23.5f Check that cam chain tensioner socket is secure

1	Input shaft	22	Retaining pin
2	Input shaft right-hand bearing	23	O-ring
3	Thrust washer	24	Final drive sprocket
4	Bearing half ring	25	Bolt – 2 off
5	Bush	26	Lock washer
6	Output shaft 5th gear pinion	27	Output shaft with bearing and 4th gear pinion
7	Splined thrust washer	28	Output shaft 6th gear pinion
8	Circlip – 2 off		
9	Output shaft 2nd and 3rd gear pinion	29	Thrust washer
10	Special thrust washer	30	Bush
11	Output shaft 6th gear pinion	31	Output shaft 2nd gear pinion
12	Splined sleeve	32	Special thrust washer
13	Splined thrust washer	33	Lock washer
14	Special locking washer	34	Output shaft 3rd gear pinion
15	Thrust washer	35	Bush
16	Output shaft 4th gear pinion	36	Circlip
17	Thrust washer	37	Output shaft 5th gear pinion
18	Output shaft left-hand bearing	38	Bush
19	Oil seal	39	Output shaft 1st gear pinion
20	O-ring	40	Thrust washer
21	Bearing retainer block		

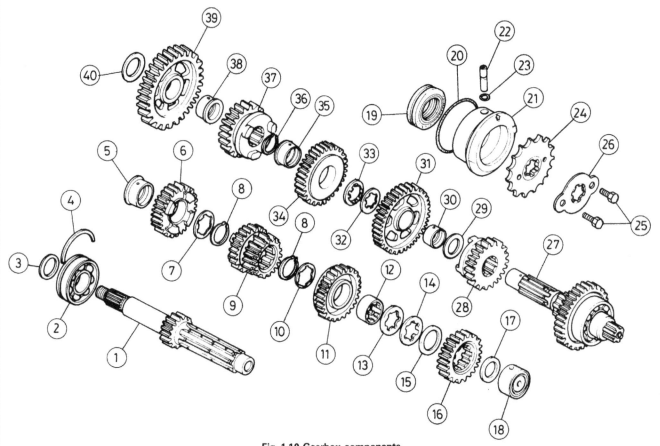

Fig. 1.10 Gearbox components

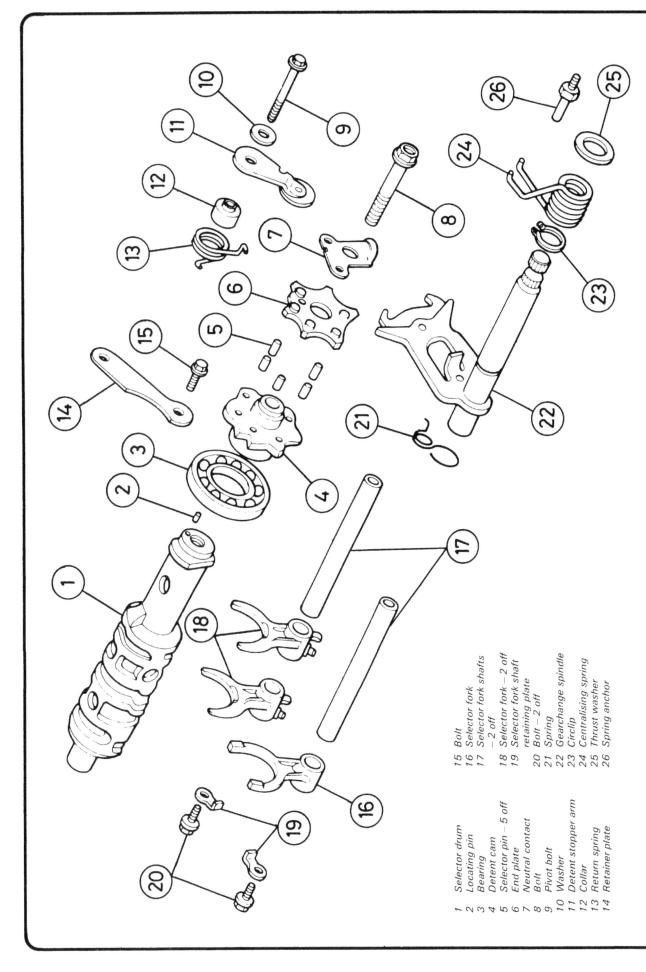

Fig. 1.11 Gearchange mechanism

1   Selector drum
2   Locating pin
3   Bearing
4   Detent cam
5   Selector pin – 5 off
6   End plate
7   Neutral contact
8   Bolt
9   Pivot bolt
10 Washer
11 Detent stopper arm
12 Collar
13 Return spring
14 Retainer plate
15 Bolt
16 Selector fork
17 Selector fork shafts – 2 off
18 Selector fork – 2 off
19 Selector fork shaft retaining plate
20 Bolt – 2 off
21 Spring
22 Gearchange spindle
23 Circlip
24 Centralising spring
25 Thrust washer
26 Spring anchor

24 Engine reassembly: joining the crankcase halves

1   Check that the two locating dowels are in place at the front edge of the crankcase upper half. Lubricate the selector fork grooves of the input shaft 5th and 6th gears and the output shaft 2nd gear with molybdenum disulphide grease. A trace of the above grease should also be applied to the crankshaft main bearing journals.

2   Apply a thin film of silicone rubber gasket compound to the jointing face of the lower half of the crankcase, taking care **not** to apply it within about 2 – 3 mm of the main bearing inserts. Apply a thin film to the upper half, noting the above precaution.

3   Lower the crankcase lower half onto the inverted upper half, guiding the selector forks into engagement, and ensuring that the oil pump pinion engages correctly. Coat the threads and the undersides of the heads of the ten main bearing bolts with molybdenum disulphide grease and run them into position; tighten them in the sequence shown in Fig.1.12 to the torque setting for the 8 mm bolts. Install the remaining lower crankcase bolts and tighten them evenly, in a criss-cross pattern to the torque setting for the 6 mm bolts.

24.3a Offer up lower crankcase half, ensuring that forks engage

Crankcase bolt torque settings

8 mm bolts	2.2 – 2.6 kgf m (16 – 19 lbf ft)
6 mm bolts	1.0 – 1.4 kgf m (7 – 10 lbf ft)

4   Turn the crankcase assembly over and fit the remaining bolts, tightening them evenly to the above torque settings.

5   Inside the sump aperture, fit the U-shaped oil pipe, the pressure relief valve and the pickup strainer. Fit new O-ring to the dowelled recess next to the strainer. Examine the large O-ring which seals the sump, renewing it if it is abraded or broken. To fit a new O-ring, place it over the sump and press it into the groove at the four corners, then work around the groove, pressing the O-ring in at regular intervals. This procedure avoids the embarassment of finding 'spare' O-ring, a common problem if working around the groove. Offer up the sump, then fit and tighten the retaining bolts.

24.3b Tighten crankcase bolts to specified settings ...

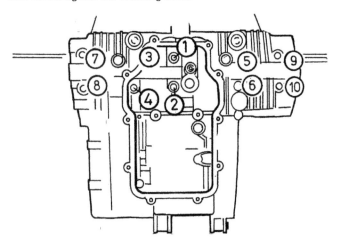
Fig. 1.12 Lower crankcase/main bearing tightening sequence

25 Engine reassembly: refitting the alternator and selector mechanism cover

1   Clean the tapered end of the crankshaft and the internal taper of the alternator. Position the alternator rotor and refit the central retaining bolt. Hold the crankshaft by the same method employed during removal, and tighten the bolt to 4.6 – 5.4 kgf m (33 – 39 lbf ft).

2   Check that the alternator stator is secure in the outer cover, and that the brushes move freely in their holders. Degrease the brushes and the alternator slip-ring. Place a new gasket on the cover, which can then be offered up and secured to the crankcase.

3   Examine the gearchange shaft oil seal in the outer cover, renewing it if worn or damaged. Lubricate the seal with engine oil and wind some electrical tape around the shaft splines to avoid scoring the seal. Fit the locating dowels and a new gasket, then fit and secure the cover.

24.3c ... noting bolts inside sump aperture

24.5a Fit U-shaped oil pipe using new O-rings ...

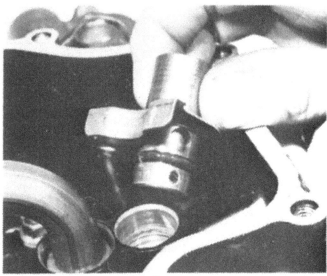

24.5b ... then fit pressure relief valve ...

24.5c ... followed by pickup strainer

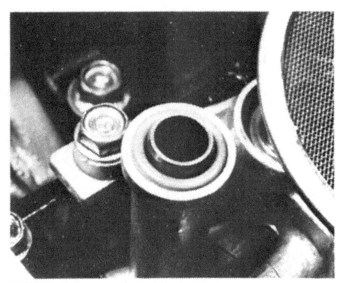

24.5d Check that pump O-ring is in place

24.5e Then fit the sump using new O-ring as required

25.1a Place alternator rotor over crankshaft taper ...

25.1b ... and tighten the retaining bolt

25.2 Fit new cover gasket and refit cover and stator assembly

25.3a Fit the selector mechanism cover using a new gasket

25.3b Refit the neutral switch lead and secure terminal nut

26 Engine reassembly: refitting the clutch

1   Place the flat anti-judder spring seat over the clutch centre, followed by the dished anti-judder spring with its convex face towards the clutch centre flange. Coat the clutch plain and friction plates with oil, then fit them over the clutch centre in an alternating sequence, starting and finishing with a friction plate. Position the external tangs of the friction plates so that they align, then place the pressure plate against the last plain plate.

2   Taking care not to disturb the alignment of the plates, invert the assembly and fit two of the springs, using the plain washers beneath the bolt heads to secure them. Fit the plate and centre assembly into the clutch drum, noting that, where necessary, the two bolts can be slackened to permit realignment of the plates.

3   Fit the Belville washer with the 'OUTSIDE' mark facing outwards, then fit the clutch centre nut. Lock the crankshaft and tighten the nut to 4.5 – 5.5 kgf m (32 -- 40 lbf ft).

4   Remove the two bolts and discard the plain washers. Lubricate the clutch release bearing, then fit the remaining springs. Place the release plate over the springs, securing it with the four retaining bolts.

5   Grease the clutch release mechanism, remembering to fit the push rod in its recess. The clutch outer cover can now be fitted, using a new gasket. Tighten the cover screws evenly and progressively to avoid any risk of warpage.

26.1a Place flat spring seat washer over clutch centre ...

26.1b ... noting that it fits inside clutch plate as shown

26.1c Fit anti-judder spring with concave face uppermost

26.1d Build up clutch plates in alternate sequence ...

26.1e ... then fit pressure plate

26.2 Fit two bolts, washers and springs to lock assembly

26.3a Fit Belville washer and clutch centre nut ...

26.3b ... and tighten to specified torque

26.4 Refit clutch release plate and lubricate bearing

26.5a Fit pushrod into cover recess ...

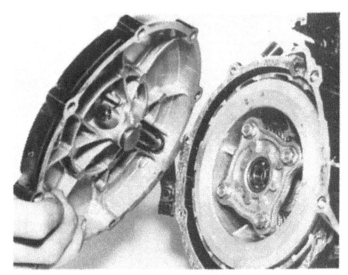

26.5b ... then install cover using a new gasket

27 Engine reassembly: refitting the ignition pick-up assembly and starter drive

1   Position the starter chain guide in the casing recess, having first positioned the ignition pickups. Each is retained by a single bolt and is located by a small dowel pin. The lead between the two pickups should be routed above the crankshaft and is held in place by a pressed steel guide plate. This, too, is secured by a single bolt.
2   Check that the starter clutch is assembled correctly and that it operates freely, then loop the chain around it and the small driving sprocket. Slide the clutch assembly over the splined end of the crankshaft, noting that the punch mark on the clutch body must align with the groove in the crankshaft end. Fit the clutch retaining bolts and tighten it to 4.8 – 5.2 kgf m (35 – 38 lbf ft).
3   Slide the starter motor extension shaft through and into engagement with the driving sprocket. Check that the wiring grommet is located in its cut-outs in the casing, then fit the outer cover.

27.1a Note that encapsulated pickup coils are located by pin

27.1b Wiring runs **behind** chain guide

27.1c Tighten pickup bolts and guide plate bolt

27.2a Check and lubricate starter clutch rollers ...

27.2b ... and install sprocket as shown

27.2c Offer up starter drive assembly ...

27.2d ... checking that notch and index mark align

27.3 Slide starter motor extension shaft into position

28 Engine reassembly: installing the pistons and cylinder block

1   Lubricate the connecting rod small-end eyes with molybdenum disulphide grease, then fit each piston in turn to its respective connecting rod. The crankcase mouths should be packed with clean rag to catch any errant circlips. Note that the 'IN' mark on each piston crown must face the inlet side (towards the rear of the crankcase). If the gudgeon pins prove stubborn, try warming each piston in very hot water to ease installation. Secure the gudgeon pins using **new** circlips.
2   Fit the dowel pins to the two holes in front of the outer cylinders, then place a new cylinder base gasket over the holding studs. Holding the cam chain taut and guiding the outer pistons into the crankcase mouths, rotate the crankshaft to bring the inner pair of pistons to TDC (top dead centre). Space the piston rings so that the end gaps lie about 120° apart and make sure that the oil rail end gaps do not coincide.
3   If working alone it is easier to use piston ring compressors to facilitate the insertion of the pistons into the bores. With an assistant, however, it is relatively easy to work the rings into the bores by hand. Position the cylinder block about $\frac{1}{2}$ inch above the pistons, then slowly lower it over them. Lubricate the bores with engine oil. As the rings reach the bores, ease each one into the tapered lead-in, taking care that the ring ends do not catch and break. Once the inner pair are installed, rotate the crankshaft to bring the outer pistons to TDC and repeat the sequence. When all four pistons are engaged, the cylinder

28.1a Fit pistons with 'IN' mark towards rear of crankcase

28.1b Fit new circlips, positioning ends away from notch

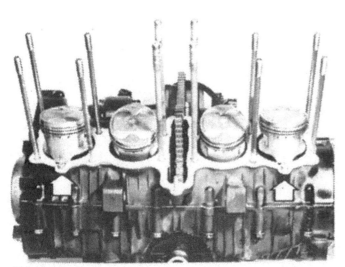

28.2 Fit dowel pins (arrowed) and new base gasket

28.3a Work block over pairs of pistons as shown

28.3b Fit and tighten base nuts

29.1 Fit chain guide in recess, then fit dowels and gasket

block can be pushed home against the base gasket, having first made sure that the cam chain has been fed up through the tunnel between the centre cylinders and secured by passing a length of wire through it. Fit the cylinder base nuts and tighten them securely.

29 Engine reassembly: installing the cylinder head

1   Fit the dowel pins in their holes to the rear of cylinders one and four, then place a new cylinder head gasket over the holding studs. Apply a thin film of jointing compound to the area of cylinder head gasket around the studs. Place the cam chain guide blade in its recess at the rear of the cam chain tunnel.
2   Pass the cam chain up through the tunnel in the cylinder head, securing it by passing a screwdriver through the chain loop. Lower the

cylinder head over the studs, checking that it seats fully. Fit the outermost eight cylinder head nuts, finger-tight only at this stage.
3   Remove the R-pin and clevis pin which secures the cam chain tensioner blade to the tensioner body assembly. Arrange the cam chain so that it passes between the forked ends of the tensioner arm. Refit the clevis pin and R-pin to secure the tensioner blade. Lower the tensioner assembly into place, making sure that the lower end of the blade engages in its recess.
4   Fit the remaining cylinder head nuts, thus securing the tensioner assembly. Refer to the accompanying photograph and tighten the cylinder head nuts in the sequence shown to the following torque settings:

Cylinder head nuts   2.0 – 2.4 kgf m (14 – 17 lbf ft)
Cylinder head/tensioner nuts   2.6 – 3.0 kgf m (19 – 22 lbf ft)

1   Cam chain
2   Camshaft sprocket – 2 off
3   Bolt – 4 off
4   Cam chain upper guide
5   Cam chain front guide
6   Cam chain tensioner
7   R-pin
8   Clevis pin
9   Tensioner blade
10 Spring sleeve
11 Spring

Fig. 1.13 Cam chain and tensioner

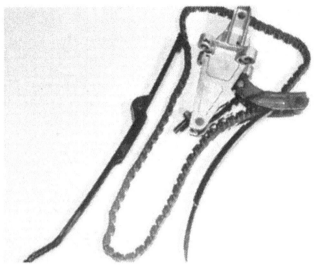

29.3a General arrangement of cam chain and tensioner parts

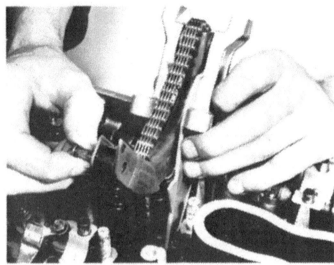

29.3b Pass chain through tensioner and fit clevis pin ...

29.3c ... securing it with its R pin

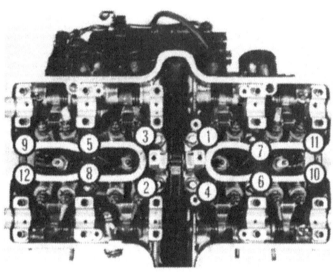

29.4a Cylinder head nut torque sequence

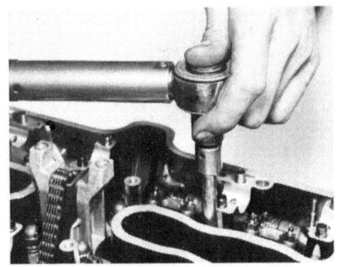

29.4b Note that a thin-walled socket is needed

30 Engine reassembly — camshaft installation and valve timing

1   Lubricate the camshaft bearing journals with molybdenum disulphide grease. Place the camshaft sprockets over the camshafts, ensuring that the timing marks face away from the camshaft flange face. Manoeuvre the camshafts into position inside the chain loop, noting that the exhaust camshaft incorporates the tachometer drive gear. Position the camshafts so that the cam lobes of cylinder No 1 (first cylinder from the left-hand side of machine) both face upwards.
2   Remove the timing inspection cap from the starter drive cover, and rotate the crankshaft until the 'T' mark aligns with the index mark on the side of the cover aperture. Fit the camshaft bearing caps, starting with the centre caps and working outwards. Tighten the cap bolts evenly and progressively so that the camshafts are pulled down squarely onto the journals. The final torque figure for these bolts is 1.0 – 1.4 kgf m (7 – 10 lbf ft). Note that the centre caps cannot be secured at this stage.
3   Arrange the cam sprockets so that the two small dots on each lie parallel to the cylinder head gasket face, then arrange the cam chain loop around them. To allow sufficient clearance while the sprockets

are positioned against the camshaft flanges it is necessary to release pressure from the tensioner by the method shown in the accompanying illustration, using a screwdriver and a length of stiff wire with a hooked end. Hold the wire taut whilst lifting the sprockets into position.

4   Recheck the crankshaft timing marks and the sprocket timing marks, then insert one of the sprocket holding bolts on each shaft, noting that Loctite or a similar locking fluid should be used on the bolt threads. Rotate the crankshaft through 360°, then fit the remaining bolts, tightening them to 2.2– 2.6 kgf m (16 – 19 lbf ft). Rotate the crankshaft through a further 360° and tighten the first pair of bolts to this figure. Align the crankshaft timing marks and recheck the timing before moving on.

5   Fit the internal oil pipes and cam chain upper guide as shown in the accompanying photographs. Prime the oil pockets beneath each cam lobe with engine oil so that the lobes will be submerged during each revolution. Check and adjust the valve clearances as described in Routine Maintenance. Clean and examine the cylinder head cover gasket, renewing it where necessary. Apply a small amount of sealant around the extended section of the gasket at each end of the camshafts, then fit and secure the cylinder head cover.

30.2a Rotate crankshaft to align 'T' mark

30.2b Fit camshaft bearing caps noting cast-in numbers

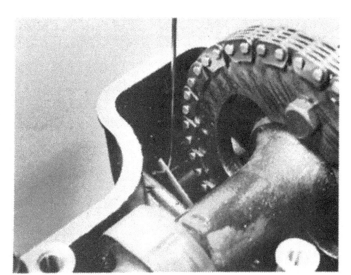

30.3a Pull here using hooked wire ...

30.3b ... and depress tensioner to release pressure

30.4 Fit cam sprockets and secure, then re-check timing

30.5a Fit internal oil pipes ...

30.5b ... and upper chain guide ...

30.5c ... then secure remaining cap bolts

30.5d Check cover O-ring and install

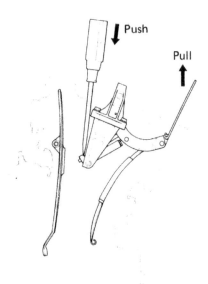

Fig. 1.14 Method of releasing cam chain tensioner pressure

Push

Pull

31 Engine reassembly: refitting the extended oil pipe, starter motor and oil pressure and neutral switch leads

1   Fit the external oil feed pipe, using new sealing washers on the unions to preclude the risk of leakage. Tighten the union bolts and oil pressure switch firmly. Manoeuvre the starter motor into position, then fit and secure the two retaining bolts which hold it to the crankcase. Refit the oil pressure switch lead, covering its terminal with the rubber boot. Reconnect the neutral switch lead, but do not tighten the terminal nut fully at this stage.

32 Engine reassembly: installing the engine/gearbox unit in the frame

1   Lift the assembled engine/transmission unit back into the frame cradle by reversing **exactly** the removal sequence. At least two people will be required for this operation and it is essential not to rush things. Note that the end caps on the rear lower mounting point may be displaced if knocked. Make certain that these are in position before lowering the unit fully home.
2   Assemble the various engine mountings, but do not tighten any of the bolts until all are in position. Do not omit the engine earth lead which is attached to the rear upper engine mounting bolt. Once all

31.1a Fit external oil feed pipe and secure oil pressure switch

31.1b The starter motor can now be slid into place

mountings are in position, tighten them to the torque figures shown below:

8 mm bolts	2.0 – 2.4 kgf m (14 – 17 lbf ft)
10 mm bolts	3.5 – 4.5 kgf m (25 – 33 lbf ft)

3   Fit the gearbox sprocket in the loop of the final drive chain and slide it over the output shaft. Fit the splined retainer, turning it until the bolts can be fitted. It is advisable to use a thread locking compound such as Loctite on these threads. Fit and secure the steel chain guide plate to the front of the sprocket, noting that the wiring should run behind it. Offer up the sprocket cover and check that the neutral switch lead dust boot is positioned correctly. Tighten the switch terminal, then secure the cover, noting that special waisted bolts are used. Fit the starter motor lead, then pull the rubber boot over the terminal to protect it.

4   Reconnect the throttle and choke cables to the carburettor bank, noting that this must be done **before** the carburettors are fitted. Manoeuvre the assembly into position between the intake and air cleaner adaptors, then ease the adaptors around the carburettor stubs. This stage is not easy and will require patience and dexterity to accomplish. Once all of the rubbers are engaged properly, secure the retaining clips. **Note:** The rubbers from the air cleaner casing are bonded to rather delicate flat diaphragms to allow normal movement when the engine is running. These are delicate and are easily torn if care is not taken when fitting the carburettors.

5   Reconnect the clutch cable and adjust it to give 10 – 20 mm free play at the end of the clutch lever. Fit the regulator/rectifier unit to the frame, then trace and reconnect the alternator wiring. Trace the ignition pickup wiring and reconnect this at the igniter unit. If it is not in position, install and secure the tachometer drive gearbox, then reconnect the tachometer cable.

6   Refit the oil cooler assembly to the frame, remembering to fit the bracket which holds the oil cooler hoses. Refit the oil cooler hoses to the sump using new sealing rings. Refit the coils to the underside of the top tube. Reconnect the low tension leads in the order in which they were removed, then connect the spark plug leads, taking note of the cylinder number markings near the plug caps. It is best to fit the oil filter assembly at this stage, before the exhaust system is in place and access is less easy. Use a new O-ring to seal the cover, and tighten the retaining bolt to 2.8 – 3.2 kgf m (20.0 – 23.0 lbf ft). **Do not** exceed this figure.

7   Fit new sealing rings to the exhaust ports, using grease to hold them in position while the exhaust system is fitted. Manoeuvre the system under the machine, and arrange the retainers as shown in the accompanying photograph, using pvc tape to hold them. Lift the assembly into position and fit the rear mounting bolts to retain it. Place each pair of split collets in position, sliding the retainers into place to hold them. Fit the retaining nuts finger tight. Once all four retainers are in position, tighten the nuts evenly to 0.9 – 1.3 kgf m (6 – 9 lbf ft). The rear mountings can now be tightened fully.

8   Check that all wiring and cables are correctly routed and clipped or taped to the frame, then refit the fairing (F2 models only, refer to

Chapter 4). Install the battery and carry out a quick electrical check to make sure none of the connectors has been accidentally separated. Refit the gearchange and brake pedals and check the operation and position of each. Top up the crankcase with approximately 2.3 litre (4.0 lmp pint) of the recommended grade of engine oil (see Routine Maintenance for further details). Refit the fuel tank and reconnect the fuel gauge sender leads, the vacuum pipe and the fuel pipe. Reconnect the crankcase breather hose and refit the seat and side panels.

33 Starting and running the rebuilt engine

1   Turn the fuel tap lever to the 'Prime' position and allow a few minutes for the float bowls to fill. Check that the engine kill switch is set to the 'Run' position, then switch on the ignition and check the warning lamps. Put the choke lever to the 'on' position and try to start the engine. This may take a few seconds while the fuel mixture stabilises, but if the engine shows no signs of firing after 10 – 20 seconds, stop and check the ignition connections.

2   There will probably be a fair amount of smoke in the exhaust during the first few minutes, due to the excess oil used during assembly burning off. Check that the oil pressure lamp goes off, then check around the engine for leaks. Stop the engine and allow it to settle for several minutes, then check the oil level.

3   Before taking the machine for a test ride, check and where necessary adjust the brakes, final drive chain, throttle cables and idle speed. Give the whole machine a visual check for loose fasteners and incorrectly routed cables.

4   The rebuilt engine will need time to settle down, even if parts have been replaced in their original order. For this reason it is highly advisable to treat the machine gently for the first few miles to ensure oil has circulated throughout the lubrication system and that any new parts fitted have begun to bed down.

5   Even greater care is necessary if the engine has been rebored or if a new crankshaft has been fitted. In the case of a rebore, the engine will have to be run-in again, as if the machine were new. This means greater use of the gearbox and a restraining hand on the throttle until at least 500 miles have been covered. There is no point in keeping to any set speed limit; the main requirement is to keep a light loading on the engine and to gradually work up performance until the 500 mile mark is reached. These recommendations can be lessened to an extent when only a new crankshaft is fitted. Experience is the best guide since it is easy to tell when an engine is running freely.

6   If at any time a lubrication failure is suspected, stop the engine immediately, and investigate the cause. If an engine is run without oil, even for a short period, irreparable engine damage is inevitable.

7   When the engine has cooled down completely after the initial run, recheck the various settings, especially the valve clearances. During the run most of the engine components will have settled into their normal working locations. Check the various oil levels, particularly that of the engine, as it may have dropped slightly now that the various passages and recesses have filled.

32.1a Check mounting rubbers and renew if perished

32.1b Manoeuvre engine unit into frame cradle

32.1c ... taking care not to displace end caps on rear mountings

32.2a Fit right-hand front brackets and bolts ...

32.2b ... and the left-hand upper ...

32.2c ... and lower mountings

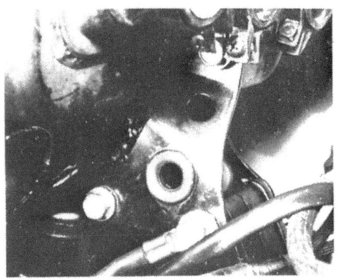

32.2d Note earth lead on upper rear mounting ...

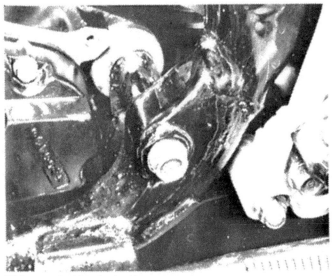

32.2e ... and rubber-bushed lower mounting

32.3a Fit gearbox sprocket and secure the retainer

32.3b Fit chain safety guide, noting wiring run

32.3c Fit cover, noting special waisted retaining bolts

32.3d Reconnect starter lead and fit the rubber boot

32.4a Hook throttle cable ends into pulley cutouts ...

32.4b ... and refit adjusters into bracket

32.4c Manoeuvre assembly back into engagement with adaptors

32.5a Fit regulator/rectifier, then reconnect alternator wiring

32.5b Plug pickup leads into underside of igniter unit

32.5c Refit tachometer drive gear and cable

32.6a Fit coil and coil bracket assembly and connect HT and low tension leads

32.6b Note wiring and cable arrangement

32.6c Refit the oil cooler connections using new O-rings

32.7a Slide new exhaust port sealing rings in place using grease

32.7b Lift assembly into position and fit the rear mountings

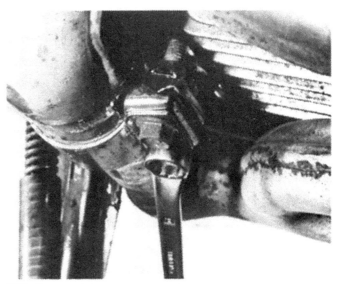

32.7c If system halves were separated, secure centre clamp

32.7d Fit collet halves and retainer ...

32.7e ... and tighten mounting nuts

32.8a Install battery and check operation of electrical system

32.8b Fit the gearchange pedal ...

32.8c ... and brake pedal, aligning index marks

32.8d Refit the crankcase breather hose

Chapter 2 Fuel system and lubrication

Contents

Specifications

Fuel tank capacity

Overall	17.0 litre (4.49/3.70 US/Imp gal)
Reserve	3.4 litre (0.90/0.70 US/Imp gal)

Carburettors

Type	4 x VE52B
Venturi size	26 mm (1.02 in)
Float level	19.2 mm (0.76 in)
Main jet	88
Slow jet	35
Pilot screw setting	$2\frac{1}{2}$ turns out
Idle speed	1200 ± 100 rpm
Fast idle speed	3000 ± 500 rpm

Lubrication system

Type	Wet sump
Capacity	3.0 litre (6.34/5.28 US/Imp pint)
Oil grade:	
Standard, all temperatures	SAE 10W-40
Above 15°C (60°F)	SAE 30
−10° to 15°C (15° to 60°F)	SAE 20
Above −10°C (15°F)	SAE 20W-50
Below 0°C (32°F)	SAE 10W
Oil pressure	5.0 kg/cm^2 (71 psi) @ 7000 rpm/80°C (176°F)

Oil pump

Type	Two-stage trochoid
Clearances (service limits):	
Inner to outer rotor	0.15 mm (0.006 in)
Body to outer rotor	0.35 mm (0.014 in)
Rotor end float	0.10 mm (0.004 in)
Pump delivery rates:	
Engine side	28 litre (36 Imp pint) per minute @ 7000 rpm
Oil cooler side	11.5 litre (20 Imp pint) per minute @ 7000 rpm

Torque settings

Component	kgf m	lbf ft
Sump drain plug	3.5 - 4.0	25 - 30
Oil filter bolt	2.8 - 3.2	20 - 23
Sump retaining bolts	1.0 - 1.4	7 - 10
Oil pressure switch	1.5 - 2.0	11 - 14

1 General description

The fuel system comprises a petrol tank from which petrol is fed by gravity to the four CD (constant depression) carburettors, via two vacuum-controlled fuel cocks. In the normal 'On' position the fuel flow from the tap is regulated by a diaphragm and plunger. The diaphragm chamber is connected by a small rubber pipe to the inlet tract, and thus will only open the plunger valve when the engine is running. Thus for normal running the fuel cock lever is set to the 'On' position at all times, although the fuel supply is turned off as soon as the engine stops. The 'Reserve' lever position provides a small emergency supply of fuel in the event that the fuel gauge is ignored, whilst the 'Prime' setting is provided to fill the carburettor float bowls should these have been dismantled for any reason or if the machine has run completely dry.

The throttle twistgrip is connected by cable to the four throttle butterfly valves. Cold starting is facilitated by providing an extra-rich cold start mixture, controlled via a cable from the handlebar-mounted 'choke' lever.

Engine lubrication is by a wet sump arrangement which is shared by the primary transmission and gearbox components. Oil from the sump is picked up by an engine-driven trochoid oil pump and delivered under pressure to the working surfaces of the engine and gearbox components.

2 Fuel tank: removal and replacement

1 Unlock and open the seat, and remove the single fixing bolt which secures the rear of the tank to the frame. Check that the tap lever is turned to 'On' or 'Res', then free the fuel pipe after displacing the retaining clip. This can be freed by squeezing together the 'ears' of the clip and sliding it along the pipe. Work the pipe off the mounting stub using a small screwdriver. Release the vacuum pipe in the same manner.

2 Lift the rear of the tank and pull it back to disengage the front mounting rubbers. Locate and separate the fuel gauge sender connector, then lift the tank clear of the frame. Store the tank away from any possible sources of fire whilst it is removed. The tank is refitted by reversing the removal sequence.

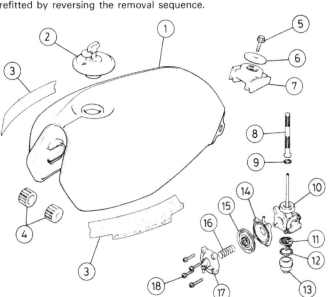

Fig. 2.1 Fuel tank and tap

1	Fuel tank	10	Fuel tap
2	Filler cap	11	Filter
3	Emblem – 2 off	12	O-ring
4	Front mounting rubbers	13	Screwed cap
5	Bolt	14	Diaphragm holder
6	Washer	15	Diaphragm
7	Damping rubber	16	Spring
8	Filter	17	Diaphragm cover
9	O-ring	18	Screw – 4 off

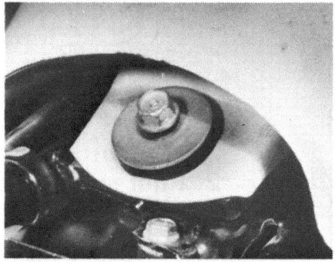

2.1a Tank is secured by rubber-bushed bolt at rear

2.1b Disconnect fuel pipe and vacuum pipe

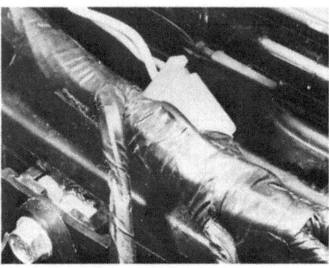

2.2 Separate the fuel level sender lead as tank is lifted

3 Fuel tap: removal, overhaul and reassembly

1   Remove the fuel tank as described in Section 2. If the tank is full, drain it into a clean metal container. Cover the workbench with soft cloth, then lay the tank on its side with the tap uppermost. Slacken the gland nut which secures the tap to the tank stub and remove the tap assembly.

2   Slacken the four screws which retain the diaphragm cover, remove the cover and then lift away the diaphragm assembly. If the diaphragm is split or holed it will be necessary to fit a new tap assembly because replacement diaphragm assemblies are not available separately.

3   Remove the screwed cap from the underside of the tap and clean out any sludge or water which may have accumulated before refitting the tap. Note that the tap lever assembly is retained by a riveted plate, and thus cannot be dismantled. Refit the tap by reversing the removal sequence.

4 Fuel and vacuum pipes: examination

1   The fuel and vacuum pipes are made from synthetic rubber and will normally require little attention for a number of years. Inspect the pipes periodically and renew any which show signs of cracking. Note that leaks from the fuel pipe can be dangerous, whilst air leaks into the vacuum pipe will render the fuel tap inoperative or may upset carburation.

2   When fitting new pipes always use synthetic rubber tubing of the correct size, **never** natural rubber which may swell and disintegrate under the action of petrol. Clear plastic fuel pipe is a tolerable substitute, but this tends to age quickly as the plasticisers are leached out.

5 Carburettors: removal and installation

1   Remove the fuel tank as described in Section 2 and pull off the side panels. Slacken the hose clips which secure the air cleaner adaptor hoses to the carburettors, then ease the air cleaner casing back. Release the mounting adaptor hose clips and pull the carburettor assembly back to disengage the stubs. Manoeuvre the assembly out to the right-hand side, then free the choke and throttle cables.

2   The carburettors are fitted by reversing the above procedure, noting that the cables should be fitted **before** the assembly is fitted to the mounting stubs. Take great care when fitting the air cleaner stubs. These are bonded to a thin and easily torn diaphragm section.

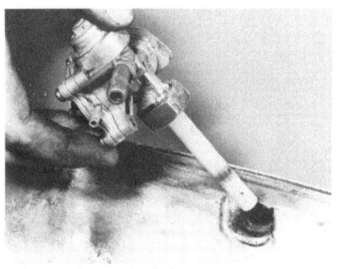
3.1 Release tap by unscrewing gland nut

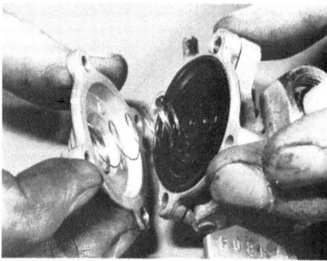
3.2a Release diaphragm cover and spring ...

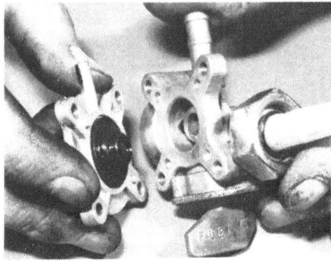
3.2b ... then remove the diaphragm assembly

3.3 Release cap to clean out any sediment or water

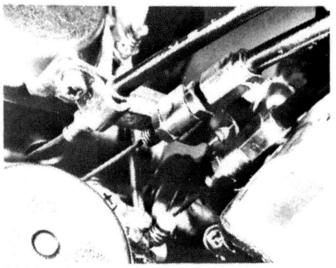

5.1 Release throttle and choke cables during removal

jet needle should be free from scoring or ridges, and must not be bent. This can be checked by rolling it on a flat surface. Note that it is preferable to renew the needle and needle jet as a pair.

4   Clean the various fuel and air jets by blowing them through with compressed air. If a stubborn blockage is encountered, use a **nylon** bristle to clear it. On no account use wire because this will damage the carefully calibrated jet drilling. The passages in the carburettor body should also be cleaned, using compressed air.

5   Check that the float is sound by shaking it. If fuel can be heard inside the float it must be renewed. The float needle valve has a tapered seating face. This should not show evidence of a ridge in its surface. When refitting the float assembly, check that the distance between the gasket face and the furthest point of the float is 19.2 mm (0.75 in) with the float arm just touching the top of the valve needle. If necessary, correct this clearance by judicious bending of the valve operating tang.

6   Assemble the carburettors by reversing the dismantling sequence. Note that the jets should be fitted firmly but not overtightened. Refit the float bowl, using a new gasket where necessary. Screw the pilot screws fully home, then back the screw out to its original setting. When fitting the valve and diaphragm assembly, ensure that the small locating tab is fitted in its recess, then fit the spring and cover.

6   Carburettors: dismantling, overhaul and reassembly

1   Slacken and remove the four screws which retain the vacuum chamber cover to each carburettor, placing each one in a marked container to indicate the instrument to which it belongs. Place all parts subsequently removed in the containers to preclude any risk of their becoming interchanged. Peel away the edge of the diaphragm, taking care not to tear it. Lift away the throttle valve and the diaphragm as an assembly. Using a socket or a box spanner, remove the hexagon-headed needle retainer and spring and displace the needle.

2   Release the four screws which retain the float bowl and lift it away. Displace the float pivot pin and lift away the float and float needle valve. Unscrew and remove the main, slow and 'by-starter' (cold start) jets (see photographs). If it is wished to remove the pilot screws, turn each one clockwise until it seats, counting the number of turns or part turns required. This will allow the original setting to be duplicated during assembly.

3   Examine the diaphragm for signs of splitting or deterioration. If renewal is required note that it will be necessary to purchase the throttle valve complete; the diaphragm is not available separately. The

6.1a Remove cover screws and lift away cover, spring and valve assembly

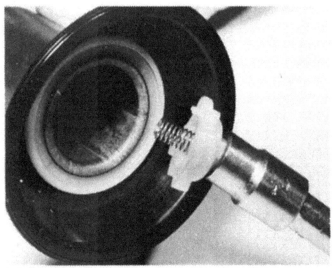

6.1b Remove plastic retainer and spring ...

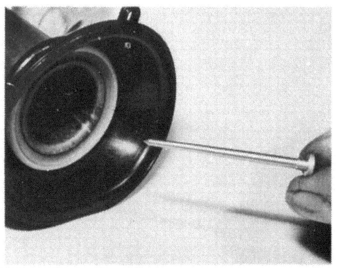

6.1c ... and displace the needle

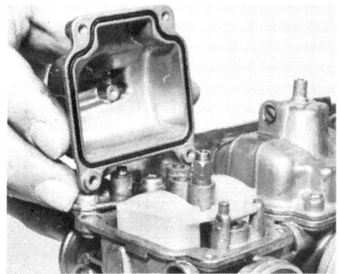

6.2a Float bowl is retained by four screws and sealed by O-ring

6.2b Remove float pivot pin and lift away float and needle

6.2c Remove main and needle jets as an assembly ...

6.2d .. then unscrew slow and cold-start jets

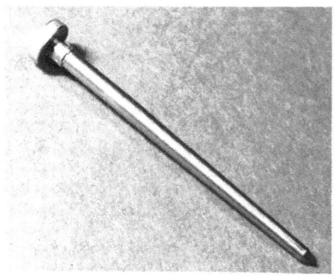

6.3 Renew needle if scored, worn or bent

6.6 Note locating tab on edge of diaphragm

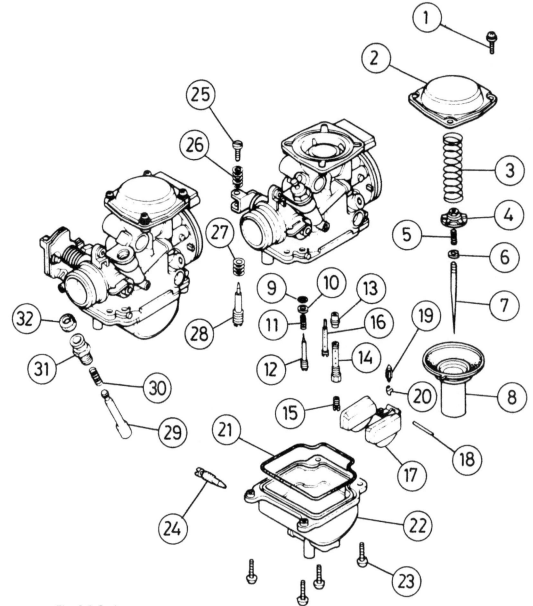

Fig. 2.2 Carburettors

1	Screw	17	Float
2	Vacuum chamber	18	Float pivot pin
3	Return spring	19	Float needle valve
4	Needle retainer	20	Needle valve clip
5	Spring	21	O-ring
6	Jet needle clip	22	Float bowl
7	Jet needle	23	Screw
8	Diaphragm and throttle valve	24	Drain screw
9	O-ring	25	Synchronising screw
10	Washer	26	Spring
11	Spring	27	Spring
12	Slow jet	28	Pilot adjusting screw
13	Needle jet	29	Throttle stop screw
14	Needle jet holder	30	Spring
15	Main jet	31	Stop screw housing
16	Cold start jet	32	Sealing ring

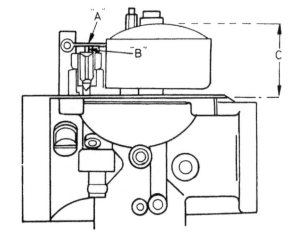

Fig. 2.3 Measuring the float height

A   Operating tang   C   19.2 mm (0.75 in)
B   Valve needle

7.2 Cold start links are locked in place by screws

7.3 Use new O-rings on connector studs and note location lugs

7 Carburettors: separation

1   Note that it is not normally necessary or desirable to separate the individual carburettors, unless it becomes necessary to attend to the fuel connection stubs, linkage or the body itself. Start by unhooking the cold start shaft and throttle shaft springs. Screw each of the synchronising screws in, counting the number of turns and part turns. Make a note of these for reference during reassembly, then slacken them to release spring tension on the linkage.
2   Slacken the cold start link locking screws and withdraw the shaft from the right-hand side. Using an impact driver, slacken and remove the sixteen mounting bracket screws and remove the brackets. Separate each carburettor, taking care not to damage the connecting stubs.
3   When assembling the carburettor bank, use new O-rings on the connecting stubs, lubricating them with a smear of oil prior to engagement. Fit the front mounting bracket, leaving the retaining screws slightly loose. The screws must be tightened with the carburettors in exact alignment. This can be achieved by placing the assembly,

with the manifold stubs uppermost, on a surface plate or a sheet of plate glass. Tighten the left-hand screw of each instrument first, followed by the right-hand screws, to ensure that the carburettors are held evenly.
4   Invert the assembly and fit the rear mounting bracket in the same manner as described above. Slide the starter shaft into positon and tighten the lock screws, then check that the mechanism operates smoothly. Hook the ends of the throttle and starter shaft springs over their locating pins. Check that the throttle linkage and springs are arranged correctly (see the accompanying line drawing).
5   Turn the synchronising screws clockwise until they seat, then unscrew them to their original settings noted during separation. Check that the distance between the throttle plate and the bypass hole in each carburettor is exactly equal. If necessary, adjust the synchronising screws to correct this. Check that the throttle mechanism operates smoothly with no signs of resistance or tight spots. Note that after the carburettor assembly has been installed it will be necessary to check and adjust carburettor synchronisation, using vacuum gauges as described below.

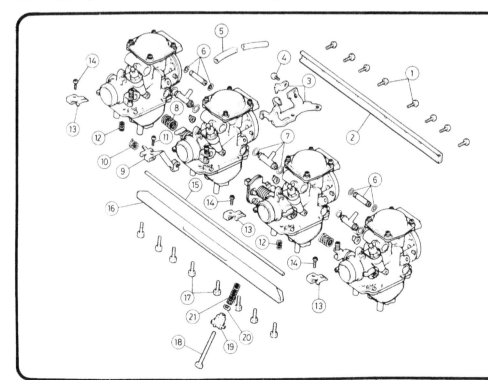

Fig. 2.4 Carburettor linkages

1   Screw – 8 off
2   Rear mounting bracket
3   Throttle cable bracket
4   Screw
5   Fuel pipe – 2 off
6   Connecting stub – 2 off
7   Connecting stub – 3 off
8   Spring – 2 off
9   Cold start lever
10   Spring
11   Screw
12   Spring – 2 off
13   Cold start link –
     3 off
14   Screw – 3 off
15   Cold start shaft
16   Front mounting bracket
17   Bolt – 8 off
18   Throttle stop screw
19   Throttle stop screw
     knob
20   Washer
21   Spring

8 Carburettors: synchronisation

1   Accurate carburettor synchronisation is essential if the engine is to run efficiently and evenly. Even a small discrepancy can cause rough running and poor fuel economy, and may even create unusual engine noises due to increased backlash in the transmission. The synchronisation procedure requires the use of a good quality vacuum gauge set designed for use on multi-cylinder motorcycle engines. These are quite an expensive investment, and many owners may prefer to leave this job to a Honda dealer. Those having access to the appropriate equipment should proceed as described below.

2   Open the seat, remove both side panels and remove the fuel tank. Plug the open end of the vacuum hose, and arrange a length of petrol pipe so that the fuel tank can be placed on a nearby bench to maintain a fuel supply during the test. Set the fuel tap to the prime position, start the engine and allow it to reach full working temperature.

3   Stop the engine, remove the vacuum take-off plugs and fit the vacuum gauge adaptors and hoses. Start the engine and adjust the idle speed to 1200 ± 100 rpm, using the central throttle stop screw. Check the reading on each gauge, noting that the actual figure is not important, but that there should be less than 30 mm (2.4 in) Hg difference between any two cylinders. Where necessary, adjust the synchronising screws to equalise the readings, noting that carburettor No 2 (second from left) is fixed and that the remaining instruments are adjusted to match it.

4   If adjustment was necessary, the idle speed will have increased. Reduce this to 1200 ± 100 rpm, and re-check the gauge readings, repeating as required until all cylinders read the same.

9 Carburettors: adjustments

1   The various jet sizes are predetermined by the factory to give the best possible combination of performance and economy; they should not normally require any alteration. The fitting of non-standard air filters or exhaust systems may, however, demand suitable modification of the jet sizes, and the advice of the retailer or accessory manufacturer should be sought. Note that such modifications may invalidate the machine's warranty.

2   The normal idle speed is set, with the engine at operating temperature, to 1200 ± 100 rpm. Adjustment is by way of a central throttle stop screw located below and between the centre carburettors. Pilot mixture adjustment is provided in the form of four mixture screws located near the bottom mounting bracket. These should never be disturbed needlessly because it is not easy to reset them without expensive exhaust analysis equipment. If the screws are ever disturbed, **always** screw them in until they seat, counting the number of turns or part-turns so that the original position can be duplicated. The nominal position for the screws is $2\frac{1}{2}$ turns out, and it is unlikely that this will require more than about $\frac{1}{2}$ turn adjustment.

3   Fuel level adjustment may occasionally be required, and is

8.3 Synchronising screw adjusts the position of one throttle to its neighbour

described in Section 6, paragraph 5. Note that an incorrect fuel level will upset carburation at all engine speeds.

10 Exhaust system: examination

1   The exhaust system requires little attention other than to check that it is secure and undamaged. Regular cleaning will postpone the inevitable need for renewal, but eventually the system will rot through from the inside. The standard system provides the optimum overall performance within legal noise levels, and is usually the best (though often not the least expensive) choice for a replacement system.

2   Non-standard systems may provide a better power output at specific engine speeds, and are often cheaper than the original type. Some pattern systems, however, can reduce power, and care must be exercised. Do not consider any system which does not include recommendations covering carburation modifications if these are required. Check also that the proposed system will fit, allowing the standard stands to be retained.

11 Air filter: examination

1   The air filter element is of the pleated paper type and is located in a moulded plastic casing beneath the dual seat. Regular maintenance will maintain fuel economy and performance, and is described in Routine Maintenance.

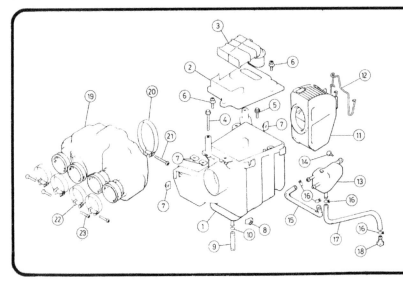

Fig. 2.5 Air filter

1 Air filter case	13 Oil/air separator
2 Cover	14 Screw
3 Air intake hose	15 Connecting pipe
4 Bolt	16 Pipe clip – 4 off
5 Bolt – 2 off	17 Drain pipe
6 Screw and washer – 2 off	18 Plug
7 Nut – 4 off	19 Air inlet box
8 Nut	20 Hose clamp
9 Drain pipe	21 Screw
10 Pipe clip	22 Hose clamp – 4 off
11 Element	23 Screw – 4 off
12 Retaining spring	

12 Engine lubrication

1   The engine shares a common lubrication supply with the gearbox and primary drive. Oil is drawn from a sump at the bottom of the crankcase and fed by a double trochoid pump to the highly-stressed plain bearing surfaces within the engine, and to the gearbox shafts, whilst the primary drive is fed by an oil spray via an internal nozzle. Full-flow filtration is provided, using a renewable paper element filter. Oil pressure is regulated by a relief valve and monitored by a pressure switch and warning lamp.
2   The CBX550 engine unit is in a fairly high state of tune, and its compact dimensions means that a limited oil reservoir can be carried. To ensure that the oil does not overheat, a separate, second stage of the oil pump circulates a proportion of the oil through a frame-mounted oil cooler radiator.

13 Oil pump: dismantling, examination and reassembly

1   The oil pump is mounted in the sump aperture. The design of the crankcase is such that it is not possible to remove the pump without first removing the engine from the frame and separating the crankcase halves, details of which will be found in Chapter 1. In view of the magnitude of the preliminary dismantling it is suggested that the condition of the pump is first checked by measuring the oil pressure.

2   With the engine at normal operating temperature (80°C, 176°F), remove the oil pressure switch and connect an oil pressure gauge to the take-off point. Check that the engine oil level is correct, then start the engine and note the pressure at 7000 rpm, comparing it with the standard pressure which is 5.0 kg/cm² (71 psi) @ 7000 rpm at 80°C (176°F). If the reading obtained is significantly less than this, remove the pump for further examination.
3   Slacken and remove the three bolts which hold the pump together. Remove the pump end covers and displace the pump rotors. Shake out the pump rotor locating pin at one end of the shaft, and remove the shaft from the pump body.
4   Examine the working faces of the rotors, the pump body and end cover faces for scoring or pitting. If this is excessive, the pump must be renewed. Assemble the pump shaft and rotors in the body, then measure the outer rotor to pump body clearances using feeler gauges. These clearances must not exceed 0.35 mm (0.014 in). Measure the clearance between the inner and outer rotor tips, again using feeler gauges. Renew the pump if the service limit of 0.15 mm (0.006 in) is exceeded. Finally, place a straight-edge across the end faces of the pump body and measure the rotor end clearance. This must not exceed 0.10 mm (0.004 in).
5   If the pump is in serviceable condition, reassemble it by reversing the dismantling sequence. Remember to fit the end cover dowels, and ensure that the inner rotors engage the driving pins. Tighten the retaining bolts and check that the pump operates smoothly before installing it and reassembling the engine.

13.3a Lift away the pump end cover ...

13.3b .. and tip out the rotors

13.3c Pin can now be displaced from shaft and removed ...

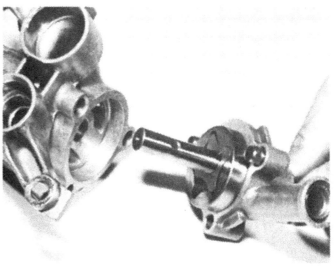

13.3d ... allowing remaining pump components to be withdrawn

13.4a Measure rotor to rotor clearance ...

13.4b ... and rotor to body clearance using feeler gauges

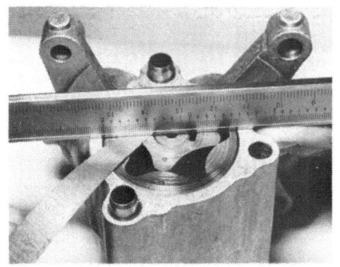

13.4c Use feeler gauge and straightedge to check end float

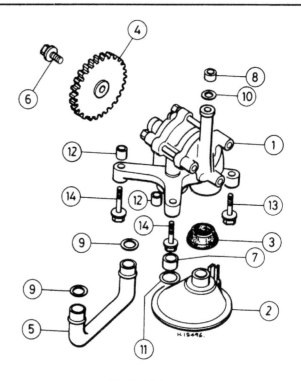

Fig. 2.6 Oil pump

1   Oil pump
2   Oil strainer
3   Sealing ring
4   Drive pinion
5   Connecting pipe
6   Bolt
7   Ring
8   Ring
9   O-ring – 2 off
10 O-ring
11 O-ring
12 Locating dowel – 2 off
13 Bolt
14 Bolt – 2 off

14 Oil filter

1   A renewable pleated paper oil filter is contained in a cast alloy filter bowl at the front of the crankcase. In the event that the filter becomes so badly choked that it can no longer pass the engine oil without imposing an excessive resistance, a bypass valve allows unfiltered oil to enter the lubrication system. Although this maintains lubrication, the unfiltered oil will cause premature engine wear, hence the need for regular oil and filter changes as described in Routine Maintenance.

15 Oil cooler: general description

1   The oil pump incorporates a pair of smaller secondary rotors which serve to pump a proportion of the oil in the sump through the oil cooler system. This is then returned to the sump where it mixes with the hot oil returning from the engine, lowering the temperature of the sump, and thus the engine. Little maintenance is required, except to check that the oil cooler matrix is kept clear of dead insects and road dirt when the machine is cleaned.

2   The oil cooler hoses are of synthetic rubber and are covered by a loose spring sheath which protects them from flying stones. Check periodically for signs of cracking of the hose material, and renew as necessary **before** a leak occurs. Note that the hose union flanges are sealed by O-rings which should be renewed together with the hoses.

Chapter 3 Ignition system

Contents

Specifications

Ignition system
Type .... Transistorised

Ignition timing
Retarded .... 15° BTDC @ 1550 ± 200 rpm
Advanced .... 37° BTDC @ 3000 ± 250 rpm

Spark plugs

	NGK	ND
Make	NGK	ND
Type:		
Standard	DR8ES	X27ESRU
Below 5°C	DR8ESL	X24ESRU
Electrode gap	0.6 – 0.7 mm (0.024 – 0.028 in)	

Pulser coil
Resistance .... 110 ± 11 ohms @ 20°C (68°F)

Ignition coil
Minimum spark gap .... 8 mm (0.3 in)

Cylinder identification .... Numbered 1, 2, 3, 4 from left-hand side of machine

---

1 General description

A fully electronic ignition system is employed, comprising two sub-systems; one for cylinders 1 and 4, and the other for cylinders 2 and 3. Each sub-system comprises a pulser or pickup coil, part of the spark unit circuitry, an ignition coil and two spark plugs. As the raised section of the ignition rotor/starter clutch passes the pickup coil a trigger pulse is generated. This signals the spark unit which sends an amplified pulse through the coil low tension windings, inducing a high tension pulse in the secondary windings. This is allowed to jump to earth across the plug electrodes.

Two sparks are produced at each plug during a complete engine cycle, but only one is used, an arrangement known as 'spare spark'. Thus when cylinder No 1 is fired, a spark is wasted in cylinder No 4 and vice versa. The spark unit is a duplex ignition amplifier serving each pair of cylinders, and incorporates the necessary electronic circuitry to advance the ignition according to engine speed. The entire system requires no maintenance other than cleaning and regapping of the spark plugs.

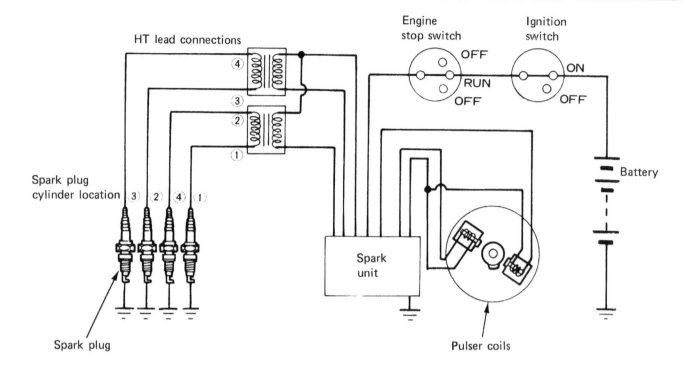

Fig. 3.1 Ignition circuit diagram

2   Electronic ignition system: testing

1   The ignition system comprises a number of sealed units making diagnosis difficult without access to expensive test equipment. In the event of a fault occurring, a number of checks can be performed to isolate the defective component, and these are described below. Follow the sequence described to minimise unnecessary duplication of the checks. The availability of a simple multimeter is assumed in many of the tests. These are available from most electrical suppliers, electronics shops and larger motorcycle dealers.

Wiring and connection test
2   Using the wiring diagram at the end of Chapter 6, trace the ignition wiring and check all connectors for disconnection, corrosion or water contamination. Clean any suspect connector and reconnect, using silicone grease inside the connector halves to prevent further water contamination. Check the wires themselves for chafing or breakages and repair as required.

Spark test
3   Remove all four spark plugs and lodge them against the engine so that the plug bodies are earthed. Some improvised brackets can be made up using stiff wire to support the plugs and to act as a sound earth if the ends are secured by some of the cover bolts or nuts. Position the plugs so that the electrodes of all four are clearly visible and then reconnect the HT leads. Switch on the ignition and check that all four plugs are sparking regularly when the engine is cranked.
4   During this test, the plugs should spark in alternate pairs; namely cylinders 1 and 4, and cylinders 2 and 3. If this is not the case, the likely faults are listed below.
5   **No spark at any plug.** Check that power is reaching the system via the ignition and engine kill switches. See Section 3 of this Chapter. Check the connections between the ignition and kill switches, the kill switch and spark unit, the spark unit and coils, the HT leads and caps and the pickup coils and spark unit. Note that it is unlikely that both sub-systems will fail simultaneously, so the most likely fault lies between the ignition switch and spark unit. If the latter is suspect it should be checked by substitution.
6   **No spark on one pair of plugs.** It can be assumed that power is

reaching the spark unit, but check the supply to the coil concerned and the condition of the coil itself as described in Section 4 of this Chapter. Check the pulser coil as described in Section 5 of this Chapter. If this fails to reveal the fault, check the spark unit by substitution.
7   **No spark at one plug only.** This is almost invariably due to a short or break in the HT lead, cap or the spark plug. The only other possible cause is an internal fault in the coil. and this can be checked as described in Section 4 of this Chapter, or by substitution.

3   Ignition and kill switches: testing

1   The following tests are conducted with a multimeter set on the 0-20V dc scale, or the nearest equivalent. A simpler test can be made using a low wattage 12V bulb with two test leads attached to it. The negative (–) test lead must be connected to a sound earth in the case of the multimeter; either lead in the case of the bulb. In each test a reading of battery voltage, or a lit bulb indicates that power is reaching that point.
2   Locate and separate the 6-pin connector from the ignition switch. Connect the positive probe to the red lead on the harness side of the connector to check that power is reaching the switch. Join the connector again, and push the probe into the back of the black (ignition) terminal. With the ignition switch on, battery voltage should be shown.
3   Locate the 9-pin connector from the right-hand switch unit. Connect the positive probe to the back of the black/white terminal and turn the ignition switch on. Check that battery voltage is indicated when the kill switch is set to the 'Run' position. Repeat this check on the black/white lead where it joins the ignition coils and at the 4-pin spark unit connector. If battery voltage is obtained, the switches and wiring are sound. If not, check back through the system and rectify the fault.
4   If a switch fault is found, try using an aerosol maintenance spray such as WD40. In an emergency, either or both switches may be bypassed with a length of wire from the battery positive terminal to the black/white terminals of the coils and spark unit. Note that this leaves no way of shutting off the ignition and thus should be repaired properly once the machine has been ridden home.

4 Ignition coils: testing

1   Check the supply to the coils using a multimeter or a test bulb arrangement as described in paragraph 1 of Section 3. Battery voltage should be present on the common black/white lead of both coils when the ignition and kill switches are on. If this is not the case, check the supply from the switches as described in Section 3.

2   To test the coils, first remove the fuel tank and trace and disconnect the coil leads at the three pin connector. Pull off the HT leads at the plug caps and release the two bolts which secure the coil assembly to the frame. Using a multimeter set on the ohms x 1 scale, check the resistance between the common black/white lead and the yellow and blue/yellow leads in turn. A sound coil should give a reading of 2.8 ohms. If either infinite or zero resistance is shown, the faulty coil must be renewed.

3   Further testing of the coil requires the use of a spark tester, and this should therefore be entrusted to a Honda dealer. For those having access to this equipment, the minimum spark gap is 8 mm (0.3 in).

5 Pulser coils: testing

1   Trace the leads from the pulser (pickup) coils to the spark unit and separate the connector. Using a multimeter on the ohmx x 1 scale, measure the resistance between the white and yellow leads (cylinders 2 and 3) and then the white and blue leads (cylinders 1 and 4). A sound coil should indicate 110 ± 11 ohms.

2   If one or both pulser coils has failed, they should be renewed.

Removal is described in Chapter 1, Section 8, installation being a straightforward reversal of this sequence.

6 Spark unit: general

1   No details of the spark unit are available, and thus it can only be checked by elimination of all other possible faults as described in the preceding Sections. If it is decided that a new unit is required, it is preferable to make a final check by substitution before purchasing the unit. In view of the cost of a new unit it may be worth trying to obtain a good secondhand item from a motorcycle breaker.

7 Ignition timing: general

1   Ignition timing is a function of the spark unit and cannot alter unless this is faulty. No maintenance is possible, but if inaccurate timing is suspected it is advisable to have the spark unit checked by a Honda dealer before engine damage occurs.

8 Spark plugs: general

1   The spark plugs form the only variable in an otherwise maintenance-free ignition system. It follows that they should be checked first in the event of an ignition fault. Maintenance is important to the continued accuracy of the ignition system, and should be carried out regularly as described in Routine Maintnenace.

4.2 Coils are attached to bracket below frame top tube

6.1 Spark unit is mounted under left-hand side panel

Chapter 4 Frame and forks

Contents

Specifications

Front forks

Type	Air-assisted oil-damped telescopic, anti-dive on left-hand leg
Oil capacity:	
Right-hand leg	292.5 – 297.5 cc
Left-hand leg	302.5 – 305.5 cc
Oil level	152 mm (5.98 in)
Oil grade	Automatic transmission fluid (ATF)
Air pressure range	0.8 – 1.2 kg/cm^2 (11 – 17 psi)
Main spring free length	480.9 mm (18.933 in)
Service limit	471.3 mm (18.555 in)
Secondary spring free length	99.0 mm (3.898 in)
Service limit	97.0 mm (3.819 in)
Fork stanchion bend service limit	0.2 mm (0.008 in)

Rear suspension unit

Type	Single hydraulically-damped coil spring unit with air assistance
Air pressure range	1.0 – 4.0 kg/cm^2 (14 – 56 psi)

Torque settings

Component	kgf m	lbf ft
Front wheel spindle	5.5 – 6.5	40 – 47
Front wheel spindle clamps	1.8 – 2.5	13 – 18
Front fork cap bolt	1.5 – 3.0	11 – 22
Handlebar mounting bolt	4.0 – 5.0	29 – 36
Upper yoke pinch bolt	0.9 – 1.3	7 – 9
Lower yoke pinch bolt	3.0 – 4.0	22 – 29
Steering stem nut	8.0 – 12.0	58 – 87
Rear wheel spindle nut	8.5 – 10.5	61 – 76
Suspension unit mounting bolt	4.0 – 5.0	29 – 36
Footrest plate Allen bolts	3.5 – 4.5	25 – 33
Suspension link to frame	4.0 – 5.0	29 – 36
Suspension link to suspension arm	5.5 – 7.0	39 – 51
Suspension arm to swinging arm	4.0 – 5.0	29 – 36
Suspension arm to suspension unit	4.0 – 5.0	29 – 36
Swinging arm pivot	8.5 – 10.5	61 – 76

1 General description

The CBX550 employs a duplex tubular steel frame of conventional construction. Front forks are of the oil-damped coil sprung telescopic type, and feature linked air assistance and an anti-dive mechanism which improves stability under braking. Rear suspension is of cantilever construction, employing a single suspension unit with air assistance. The suspension linkage, known as Pro-Link, provides a rising rate suspension. The F2 version is equipped with an integral sports fairing carried on a tubular sub-frame.

2 Front forks: removal and replacement

1 Remove the front wheel as described in Chapter 5, Section 2, tying the brake assemblies well clear of the forks. On F2 versions, remove the fairing as described in Section 11 of this Chapter.

2 Remove the fork air valve cap and release fork air pressure by depressing the valve core with a small screwdriver or a piece of wire. Once air pressure has been relieved, slacken and release the air hose union on the right-hand fork leg. The hose can then be manoeuvred clear of the steering head and unscrewed from the left-hand fork. Prise off the wire circlip which locates the handlebar on the stanchion. Note that a similar circlip is located below the handlebar, and this must be prised off once the handlebar has been removed. Slacken the pinch bolt which secures each handlebar section and pull it clear of the steering head area. Tie the handlebars clear of the forks. Remove the bottom yoke cover, and disconnect the speedometer cable from the brake plate.

3 Prise off the black plastic caps which conceal the fork brace screws. Remove the screws and lift away the brace. Remove the four

bolts which retain the front mudguard and lift it clear of the forks. If the forks are to be dismantled after removal, slacken the fork cap nuts. These can be gripped with self-locking pliers, using rag to prevent damage to the anodised finish. Slacken the upper and lower yoke pinch bolts, then remove each fork leg by pulling it downwards and twisting until it is clear of the yokes.

4   When refitting the forks, slide each leg into position and temporarily tighten the pinch bolts. After checking that the fork oil level is correct (see Section 4) tighten the cap nuts firmly. If some method of using a torque wrench can be found, the correct figure is 1.5 – 3.0 kgf m (11 – 22 lbf ft). Slacken the pinch bolts and arrange each leg so that the air valve holes are angled outwards by 45°, and the lower circlip on the stanchion is flush with the surface of the top yoke. Tighten the pinch bolts to the following torque settings. The lower yoke pinch bolts should be tightened first to 1.8 – 2.5 kgf m, then the upper yoke pinch bolts can be tightened to 0.9 – 1.3 kgf m (7 – 9 lbf ft). Refit the handlebars and locate the upper circlip. Tighten the pinch bolts to 4.0 – 5.0 kgf m (29 – 36 lbf ft).

5   Refit the air hose, routing it around the instrument drive cables. Use new O-rings on the unions and grease them during assembly. Refit the fork brace, front mudguard and front wheel. Check that there is no weight on the front wheel, then set the fork air pressure to 0.8 – 1.2 kg cm^2 (11 – 17 psi) using a fork air pump. With the stand retracted, apply the front brake and push up and down on the forks, then re-check the air pressure setting.

2.2a Release air pressure at valve (arrowed) then remove air hose union

2.2b Hose can be unscrewed from l.h. fork and removed

2.3a Fork brace screws are hidden by small plastic caps

2.3b Slacken upper pinch bolts ...

2.3c ... and lower pinch bolts ...

2.3d ... then pull fork legs down and clear of yokes

3   Front yokes and steering head bearings: removal and replacement

1   Remove the front forks as described in Section 2. On F2 machines, release the fairing bracket fasteners and remove the subframe. On all other machines, release the headlamp retaining screw, then lift out and disconnect the headlamp. Separate the wiring connectors and push them clear of the shell, then release the shell mounting bolts and lift the shell away.

2   Disconnect the instrument drive cables from the underside of the instrument panel. Remove the instrument panel retaining bolts and lift away the panel. On unfaired machines, remove the headlamp mounting bracket. Using a socket or box spanner, slacken and remove the steering stem nut. The top yoke can now be tapped upwards and removed.
3   Place an old blanket or some large rags beneath the bottom yoke to catch any steering head balls which drop free. Slacken the steering stem adjuster, holding the bottom yoke in position while it is removed. Carefully lower the yoke clear of the steering head and place it to one side. Collect the steering head balls which can now be degreased prior to examination.
4   Both the races and the balls must be free from pitting or indentation. If any of the balls are marked they should be renewed as a set. Note that apparently minor blemishes in the races can cause handling defects, so renew them unless in perfect condition. The bearing cups can be driven out of the steering head using a long drift. The lower cone can be driven off the steering stem using a cold chisel as a wedge.
5   Fit a new dust seal over the steering stem, then drive the lower cone into position using a long tubular drift. Ensure that it seats squarely and fully. A large socket can be used to drive the bearing cups into the steering head. Coat both the upper cup and lower cone with general purpose grease and use this to hold the bearing balls in place. There should be 18 upper balls and 19 lower balls. Offer up the lower yoke and retain it with the slotted nut. Tighten the nut until it seats, then back it off by $\frac{1}{8}$ of a turn. Check that the steering stem turns easily and smoothly, and that no discernible play is evident, and make any minor adjustments necessary.
6   Temporarily refit the fork legs, holding them in place with the lower yoke pinch bolts. Refit the top yoke, locating it over the fork stanchions, then tighten the top nut to 8.0 – 12.0 kgf m (58 – 87 lbf ft). Continue reassembly by reversing the dismantling sequence.

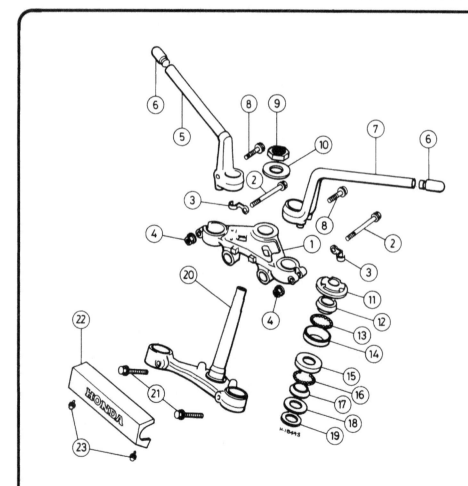

Fig. 4.1 Steering head assembly

1   Upper yoke
2   Bolt – 2 off
3   Cable guide – 2 off
4   Nut – 2 off
5   Right-hand handlebar
6   Handlebar end – 2 off
7   Left-hand handlebar
8   Bolt – 2 off
9   Nut
10   Washer
11   Adjuster nut
12   Upper bearing cone
13   Upper bearing balls
14   Upper bearing cup
15   Lower bearing cup
16   Lower bearing balls
17   Lower bearing cone
18   Dust seal
19   Washer
20   Lower yoke
21   Bolt – 2 off
22   Lower yoke cover
23   Screw and washer –
     2 off

4   Front forks legs: dismantling, renovation and reassembly

1   Remove the fork legs as described in Section 2, remembering to slacken the top bolts before removal. Strip and rebuild one fork leg at a time, and keep all internal components free from dust.

2   Remove the top bolt, noting that it will be under slight spring pressure. Remove the short spring, spring seat and the main fork spring, then invert the fork leg and expel the fork oil by 'pumping' the fork leg. Clamp the lower leg horizontally in a vice, using rag to protect the paint finish. On no account overtighten the vice or the lower leg will be distorted. Using an Allen key, remove the damper bolt from the underside of the lower leg. Note that it may be necessary to temporarily refit the fork springs and top bolt to prevent the damper rod from turning inside the lower leg.

3   Slide the dust seal away from the top of the lower leg and remove the circlip which retains the seal and top bush. Pull the stanchion sharply outwards until the bush and seal are displaced and the stanchion assembly comes out of the lower leg. Slide the seal, backing ring and top bush off the stanchion, leaving the bottom bush undisturbed unless it is to be renewed.

4   In the case of the left-hand fork leg only, release the wire circlip which retains the spring seat, spring and oil lock valve on the damper rod. Free the second wire circlip, then invert the stanchion and tip out the damper rod. Displace the damper rod seat from the bottom of the lower leg, noting that it is held in position by an O-ring.

5   Measure the free length of the main and secondary fork springs, renewing the springs in **both** legs if any have compressed to or beyond the service limits. Examine the stanchion, damper rod and lower leg for signs of wear or scoring of the sliding surfaces. Check the rebound spring and damper piston ring for wear or damage, renewing any of the above parts which are unserviceable.

6   Place the stanchion between V-blocks, and measure any bend using a dial gauge. Note that the actual bend will be $\frac{1}{2}$ of the total indicated on the gauge. Very slight bending can be corrected by skilled use of a press, and a local engineering company or motorcycle dealer may be able to carry out this work. Do bear in mind, however, that impact damage may have fatigued the metal, so if in any doubt, play safe and renew the stanchion.

7   Examine the fork bushes for wear or scoring. It is not possible to assess wear by direct measurement, but if there are obvious areas of abrasion or if play can be felt in the assembled fork leg, renew the bushes. In the case of the top bush, check that the backing ring is not deformed, and renew this as required.

8   If attention to the anti-dive unit is required, refer to Section 5 before commencing reassembly. Clean each of the fork components with methylated spirit and lay them out on a clean surface. Fit the rebound spring over the damper rod and drop it into the stanchion. In the case of the left-hand fork leg, fit the circlip, spring seat, spring, lock valve and second circlip to the damper rod. Check, and if necessary renew, the O-ring on the damper rod seat and place it over the end of the damper rod. Temporarily refit the fork springs and top bolt to retain the damper, then lower the lower leg over the stanchion assembly.

9   Coat the damper bolt threads with non-hardening locking fluid, then secure the bolt, tightening it to 1.5 – 2.5 kgf m (11 – 18 lbf ft). Remove the top bolt and fork springs. Coat the fork seal in ATF (automatic transmission fluid) and press the bush, backing ring and seal home over the stanchion and into the lower leg recess, using a length of tubing. Secure the assembly with the circlip and fit the dust seal.

10   Compress the assembled fork fully and stand it vertically. Top up with ATF to within 152 mm (5.98 in) of the top of the stanchion. Note that the quantity varies between the two legs, but that the level is identical. Refit the springs and spring seat, noting that the tapered coils should face downward, and install the top bolt. Refit the fork legs, remembering to tighten fully the top bolts.

4.2 Damper rod is secured by Allen bolt

4.3a Slide dust seal clear of lower leg ...

4.3b ... and remove circlip. Seal will be displaced as fork leg is removed ...

4.3c ... followed by lipped washer and bush

4.4a Release circlip at bottom of damper rod ...

4.4b .. to free the oil lock valve assembly

4.4c Damper rod seat is held by O-ring in lower leg

4.8a Fit rebound spring over damper rod and install in stanchion

4.8b Refit damper rod assembly as shown (l.h. leg)

4.10 Top up to prescribe oil level, using ATF

1 Air valve cap
2 Air valve
3 O-ring
4 Right-hand cap bolt
5 Left-hand cap bolt
6 O-ring – 2 off
7 Short spring – 2 off
8 Spring seat – 2 off
9 Main spring – 2 off
10 Damper rod piston ring – 3 off
11 Stanchion – 2 off
12 Damper rod – 2 off
13 Rebound spring – 2 off
14 Circlip
15 Washer
16 Spring
17 Oil lock valve
18 Circlip
19 Damper rod seat
20 O-ring
21 Circlip
22 Bottom bush – 2 off
23 Damper rod seat
24 Dust seal – 2 off
25 Circlip – 2 off
26 Oil seal – 2 off
27 Backing ring – 2 off
28 Top bush – 2 off
29 Left-hand lower leg
30 Emblem – 2 off
31 Right-hand lower leg
32 Drain bolt – 2 off
33 Sealing washer – 2 off
34 Sticker
35 Damper bolt – 2 off
36 Sealing washer – 2 off
37 Stud – 2 off
38 Wheel spindle clamp
39 Washer – 2 off
40 Spring washer – 2 off
41 Nut – 2 off
42 Allen bolt – 4 off
43 Anti-dive case
44 Spring
45 Piston
46 O-ring
47 O-ring – 2 off
48 Air hose
49 Hose union

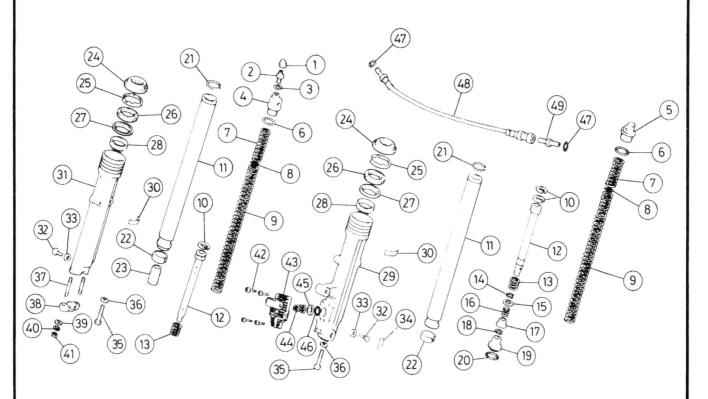

Fig. 4.2 Front forks

5 Anti-dive mechanism – removal, renovation and reassembly

1   The anti-dive mechanism can be dealt with when the front forks and wheel are in position, but because it is necessary to drain the fork prior to its removal, and to top up the fork afterwards, it is advantageous to carry out this operation in conjunction with a fork overhaul. If working with the wheel in place, release the large Allen bolt which anchors the torque link to the brake plate.

2   Remove the torque link from the anti-dive torque arm, then remove the two Allen bolts which retain the torque arm and lift it away. Remove the four small Allen bolts which retain the anti-dive case to the fork leg and lift it away. Remove the large O-ring, the piston and the spring. From the underside of the case, remove the screw, spring and check-valve ball. Remove the two screws which secure the orifice selector retainer, then remove the retainer and the selector.

3   Clean and examine each component, rejecting any that shows signs of significant wear or damage. Clean the drillings by washing the case in solvent and blowing it through with compressed air. Re-assemble the anti-dive case components, ensuring absolute cleanliness, and lubricating the piston and O-ring with ATF (automatic transmission fluid). Check that the piston moves smoothly. Refit the case assembly, tightening the holding bolts to 0.6 – 0.9 kgf m (4.3 – 6.5 lbf ft). Note that a locking compound, such as Loctite, should be applied to the screw and bolt threads.

4   Before the torque arm is installed, displace and clean its pivot sleeve and examine the dust seals. Lubricate both with silicone grease prior to assembly, and take great care not to damage the seal lips. Repeat this check on the upper pivot sleeve and seals. Note that excessive play in either pivot will impair the operation of the anti-dive mechanism.

6 Frame: examination and renovation

1   The frame is unlikely to require attention unless accident damage has occurred. In some cases, renewal of the frame is the only satisfactory remedy if the frame is badly out of alignment. Only a few frame specialists have the jigs and mandrels necessary for resetting the frame to the required standard of accuracy, and even then there is no easy means of assessing to what extent the frame may have been overstressed.

2   After the machine has covered a considerable mileage, it is advisable to examine the frame closely for signs of cracking or splitting at the welded joints. Rust corrosion can also cause weakness at these joints. Minor damage can be repaired by welding or brazing, depending on the extent and nature of the damage.

3   Remember that a frame which is out of alignment will cause handling problems and may even promote 'speed wobbles'. If mis-alignment is suspected, as a result of an accident, it will be necessary to strip the machine completely so that the frame can be checked, and if necessary, renewed.

5.2a Remove torque arm from anti-dive case

5.2b Anti-dive case can now be released from fork

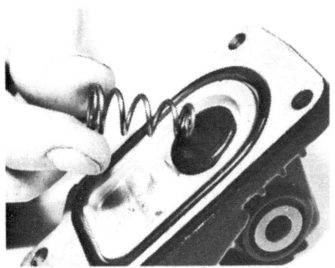

5.2c Remove the tapered spring ...

5.2d ... and the piston

5.2e Remove screw, and shake out spring ...

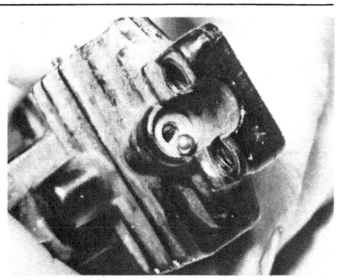
5.2f ... and check valve ball

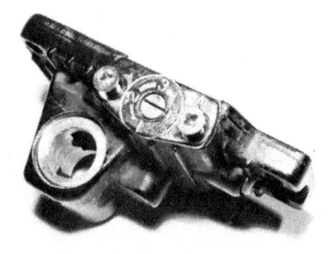
5.2g Selector retainer is secured by countersunk screws

5.4 Examine and grease sleeves during reassembly

7   Rear suspension assembly: removal, renovation and re-assembly

1   Place the machine on its centre stand and remove the rear wheel as described in Chapter 5, Section 5. Remove the final drive chain guard from the swinging arm. Slacken and remove the two Allen-headed bolts which retain the ends of the suspension arm to the swinging arm. Release the two swinging arm pivot bolts, then pull the swinging arm back and disengage it from the final drive chain.

2   Remove the two bolts which secure the suspension links to the suspension arm. The links can be pivoted clear of the arm for now. Release the suspension unit lower mounting bolt, and lift away the suspension arm.

3   To release the links from the frame, first remove the rear footrest support plates, then slacken the special bolts and nuts which secure the links. The suspension unit can be removed as required by removing its upper mounting bolt. Note the small drain hose from the suspension unit which is connected to a stub clipped to the left-hand suspension link.

4   The pivot points of the suspension arm and links take the form of sleeves carried in renewable bushes, the assembly being protected by a seal at each end. These should be degreased and inspected for wear

or scoring. The bushes are pressed into their respective bores, and must not be disturbed unless renewal proves necessary, in which case the old bushes can be driven out using a stepped drift. If care is taken, the new bushes can be inserted using a long bolt and nut and suitable spacers and plain washers. Note that the bushes must be drawn squarely into place and should be positioned so that an equal amount protrudes on each side of the casting to which they are fitted.

5   Examine the lips of the grease seals and renew any that are split, scored or distorted. Lubricate the bushes and the seals using molybdenum disulphide grease prior to reassembly.

6   Free play in the swinging arm will necessitate renewal of the bearings and seals. Note that the seals and the needle roller bearing on the left-hand side of the swinging arm will be destroyed during removal, and these should not be disturbed unless new replacements are to hand. Start by removing the short mudguard extension and the drive chain guide from the swinging arm.

7   Prise out and discard the oil seal from the left-hand boss, then drive the needle roller bearing out from the inner face, using a short drift. Alternatively, use the Honda bearing remover, Part Number 07931-MA70000 or a similar bearing extractor to pull the bearing out.

8   From the right-hand boss, remove the collar and oil seal, then release the internal circlip which locates the journal ball bearings. The bearings can now be driven out as described above. If the bearings are

in good condition and care is taken during removal, they can be re-used.

9   When refitting the bearings, pack them with grease, and grease the new oil seal. Drive the bearings squarely into the boss and refit the circlip. Fit the new seal, then place the collar into position.

10 Grease the new needle roller bearing and drive it home using Honda tool 07946-KA50000 or a suitable stepped drift. Grease and fit the oil seal, then refit the chain guide and the mudguard extension.

11 Offer up the suspension links, noting that the left-hand link carries the drain tube stub and clip, and fit the special bolts, followed by the footrest plate. Grease the suspension unit pivot sleeves and fit the end caps. Offer up the unit and fit the top mounting bolt. Fit the suspension arm to the lower suspension unit mounting, then reconnect the suspension links and pivot bolts.

12 Offer up the swinging arm, remembering to engage the drive chain loop, and fit the pivot bolts. Fit the suspension arm ends into their recesses in the swinging arm, and fit the Allen-headed pivot bolts. Check that all fasteners are secured to their correct torque settings (see Specifications), then complete reassembly by refitting the rear wheel. Finally, reconnect the suspension unit drain hose to its stub on the left-hand link.

7.1a Remove Allen bolts to free suspension arm pivots

7.1b Slacken and remove swinging arm pivot bolts ...

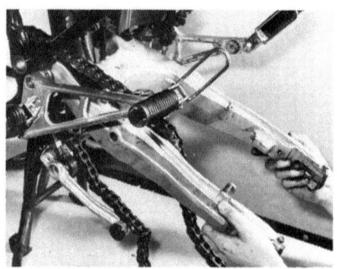
7.1c ... and manoeuvre swinging arm clear of frame

7.2a Remove the suspension arm pivot bolts ...

7.2b ... and lower suspension unit bolt to free suspension arm

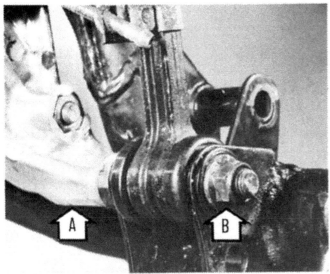

7.3 Remove footrest bracket (A) and nut (B) to free links

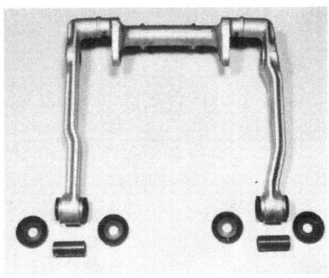

7.4 Displace and examine pivot sleeves, then grease during installation

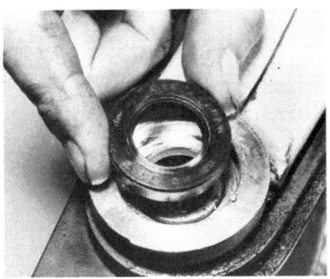

7.7a Prise out and discard the old grease seal ...

7.7b ... then drive out the worn needle roller bearing

7.8a Prise out the inner grease seal ...

7.8b ... and outer grease seal from right-hand side

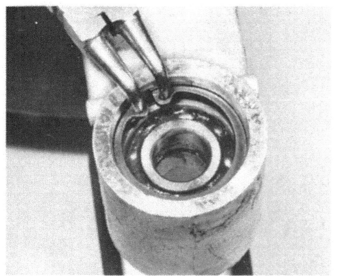

7.8c Remove the internal circlip ...

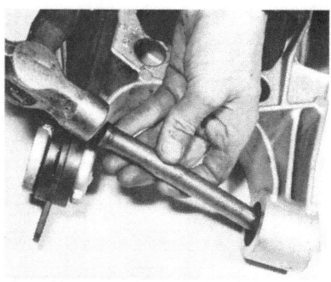

7.8d ... and drive out the bearing

7.11a Grease suspension unit top bush ...

7.11b ... fit O-rings and caps ...

7.11c ... and insert the pivot sleeve

7.11d Suspension unit can now be installed (engine shown removed for clarity)

7.12 Reconnect drain hose to stub on suspension link

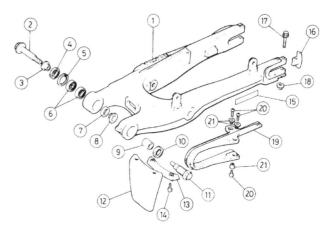

Fig. 4.4 Swinging arm

1	Swinging arm	12	Short mudguard extension
2	Right-hand pivot bolt	13	Mudguard extension plate
3	Collar	14	Bolt – 2 off
4	Oil seal	15	Emblem
5	Circlip	16	End stop – 2 off
6	Bearings	17	Bolt – 2 off
7	Oil seal	18	Nut – 2 off
8	Collar	19	Final drive chain guide
9	Needle roller bearing	20	Screw – 3 off
10	Oil seal	21	Collar – 3 off
11	Left-hand pivot bolt		

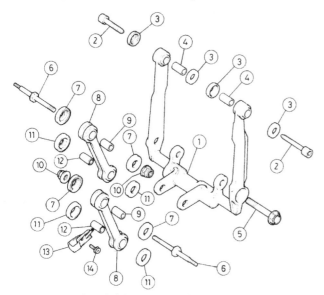

Fig. 4.3 Rear suspension linkage

1	Suspension arm	9	Upper link bush – 2 off
2	Allen bolt – 2 off	10	Nut – 2 off
3	Dust seal – 4 off	11	Dust seal – 4 off
4	Bush – 2 off	12	Lower link bush
5	Bolt – 2 off		– 2 off
6	Special bolt – 2 off	13	Drain tube clamp
7	Dust seal – 4 off	14	Screw
8	Suspension link – 2 off		

8 Rear suspension unit: removal, renovation and installation

1   The rear suspension unit can be removed as described in Section 7, or without dismantling the rest of the suspension linkage. Start by removing the Allen-headed pivot bolts which secure the suspension arm to the swinging arm via the cutouts in the footrest plate. Remove the upper suspension mounting bolt, then support the unit while the lower bolt is freed. The unit can now be manoeuvred clear of the frame.

2   The suspension unit cannot be rebuilt, but it is possible to renew the oil seal and guide bush provided that the damper rod surface is not scored or pitted. This operation requires the use of an hydraulic press and a selection of special tools, and thus is best entrusted to a Honda dealer.

3   Clean the upper mounting sleeve and coat it with molybdenum disulphide grease, then stick the seals in place with grease. Refit the unit by reversing the above sequence, tightening the various bolts to their recommended torque settings.

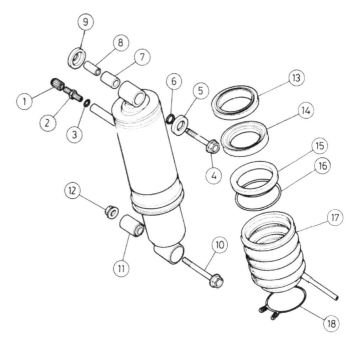

Fig. 4.5 Rear suspension unit

1	Air valve cap	10	Lower mounting bolt
2	Air valve	11	Lower bush
3	O-ring	12	Nut
4	Upper mounting bolt	13	Guide bush
5	Dust seal	14	Oil seal
6	O-ring – 2 off	15	Backing ring
7	Bush	16	Retaining ring
8	Sleeve	17	Gaiter
9	Dust seal	18	Gaiter clip

9 Stands: examination

1   The machine is equipped with a main or centre stand and also a prop stand mounted on the left-hand side of the machine. Each pivots on a bolt and nut arrangement which passes through lugs on the frame, the stands being retracted by coil springs. When cleaning the machine, clean and lubricate the stand pivots and check that the return springs are in sound condition.

10 Footrests: examination

1   The footrests are of the folding type, and are bolted to a light alloy support plate on each side of the machine. Little maintenance is required except regular lubrication of the pivot. The footrests are designed to fold if the machine is dropped, but if more extensive damage is incurred they may be unbolted for renewal.

9.1 Check stand spring condition and lubricate pivots

11 Fairing: removal and replacement – CBX550 F2

1   The F2 model features a sports fairing mounted on a tubular subframe. Removal will be necessary to gain access to the steering head area and also during engine removal, or in the event of accident damage.

2   Remove the two domed nuts which retain the fairing-mounted mirrors, and lift the mirrors away. Trace and separate the headlamp and parking lamp wiring. The fairing moulding is secured to its subframe by six bolts and two screws, the bolt heads being covered by small plastic caps which can be prised off. With the help of an assistant, remove these and lift the fairing away from the frame.

3   Disconnect the turn signal wiring, then remove the turn signal lamps from the frame. Release the clips which secure the wiring to the subframe. Remove the two extended studs which retain the subframe, followed by the single bolt which secures it to the headstock. The subframe can now be lifted away.

4   If the windscreen is to be removed for renewal, prise off the black plastic caps which cover the retaining screws, remove the screws and lift away the windscreen. When fitting the windscreen, check that it seats correctly against the weather seal, and that the headed spacers are refitted.

11.2 Do not lose special flanged nuts and rubber washers

Chapter 5 Wheels, brakes and tyres

Contents

Specifications

Tyres

Type	Tubeless
Size:	
Front	3.60H18-4PR
Rear	4.10H18-4PR
Pressures (cold):	
Front - solo	32 psi
Front - with passenger	32 psi
Rear - solo	32 psi
Rear - with passenger	40 psi

Wheels

Type	'Comstar' composite construction with aluminium alloy rims and spokes

Brakes

Type:	
Front	Twin hydraulic disc brakes
Rear	Single hydraulic disc brake
Configuration	Inboard ventilated disc, inverted twin-piston calipers
Disc thickness	11.0 - 11.2 mm (0.433 - 0.441 in)
Service limit	10.0 mm (0.394 in)
Disc runout (maximum)	0.3 mm (0.012 in)
Master cylinder bore diameter	15.870 - 15.913 mm (0.6248 - 0.6265 in)
Service limit	15.925 mm (0.6270 in)
Master cylinder piston diameter	15.827 - 15.854 mm (0.6231 - 0.6242 in)
Service limit	15.815 mm (0.6226 in)
Caliper piston diameter	31.948 - 31.998 mm (1.2578 - 1.2598 in)
Service limit	31.940 mm (1.2575 in)
Caliper bore diameter	32.030 - 32.080 mm (1.2610 - 1.2630 in)
Service limit	32.090 mm (1.2634 in)

Torque settings

Component	kgf m	lbf ft
Front wheel spindle	5.5 - 6.5	40 - 47
Front wheel spindle clamps	1.8 - 2.5	13 - 18
Rear wheel spindle nut	8.5 - 10.5	61 - 76
Brake bleed valves	0.4 - 0.7	3 - 5
Pad locating pin	1.5 - 2.0	11 - 14
Anti-dive link bolts	2.2 - 2.8	16 - 18
Brake hose union bolts	2.5 - 3.5	18 - 25
Rear master cylinder adjuster	1.5 - 2.0	11 - 14

1 General description

The wheels are of composite construction, employing light alloy rims riveted to three paired alloy spokes. At the centre the spoke sections are bolted through the brake hubs. Despite appearance, the wheel assemblies cannot be rebuilt, and no attempt should be made to dismantle them.

The brakes are of unusual design, the ventilated cast iron discs being located on three pillars around their outer edges and being capable of some axial movement. The calipers are rigidly mounted on a cast brake plate and engage over the inner edge of the discs. Each disc has one caliper, these being of twin-piston, single-sided design. The brake plate and a removable outer cover form a complete ventilated enclosure for each brake assembly, the external appearance being similar to that of a conventional drum brake. The front unit is of twin disc design, whilst rear braking is provided by a single unit.

The tyres are of tubeless construction, both being 18 inches in diameter.

2 Wheels: examination and renovation

1 The wheels require little maintenance apart from regular cleaning. In the event of suspected impact damage, check the wheel for distortion by arranging it so that it is clear of the ground. Using a dial gauge mounted on the front fork or the swinging arm, as appropriate, check the axial (side-to-side) play and radial play (ovality). In each case this should not exceed 2.0 mm (0.08 in). If distorted beyond this amount the wheel must be renewed. Note that it cannot be repaired or trued.

2 Check for bearing wear by turning the wheel and by rocking it from side to side. Any 'grittiness' or unevenness in its movement, or any discernible free play in the bearing is indicative of the need for renewal. In the case of the rear wheel, take care not to confuse wheel bearing free play with movement in the swinging arm.

3 Give the whole wheel a close visual check for dents or cracks. Any defect found will warrant immediate renewal of the wheel. Remember that a damaged rim may allow air loss from the tubeless tyre, and that this could be sudden and disastrous if signs of damage are ignored. Also, a slightly damaged wheel can suffer sudden failure due to the high stress induced in the material by small nicks or cracks.

3 Front wheel: removal and replacement

1 Place the machine on its centre stand, using blocks to raise the wheel clear of the ground. Disconnect the speedometer cable and lodge it clear of the wheel. Remove the 8 mm Allen bolts and displace and remove the caliper torque link (left-hand side). Remove the single large bolt which secures the caliper to the right-hand lower leg. Remove the three domed nuts which secure the brake disc shrouds to each side of the wheel.

2 Slacken the wheel spindle clamp nuts and remove the wheel spindle. Lower the wheel, then manoeuvre the shrouds clear of the wheel and up the fork lower legs, tying them clear of the brake area. Move the wheel as far down and forward as the brake hoses will allow, then disengage the brake caliper, brake plate and disc as an assembly. This may require judicious use of a screwdriver to ease the disc off its support. Repeat this stage on the remaining disc, then simultaneously move the wheel forward and clear of the forks while the brake assemblies are disengaged and tied clear of the forks. **Do not** leave the brake assemblies hanging on their hydraulic hoses.

3 When refitting the wheel, put a **thin** smear of Copaslip or molybdenum disulphide grease on the disc support posts and check that the friction springs are in position. Continue reassembly by reversing the removal sequence, noting that it is important to ensure that the speedometer drive engages properly. Tighten all fasteners to the torque figures shown in the Specifications.

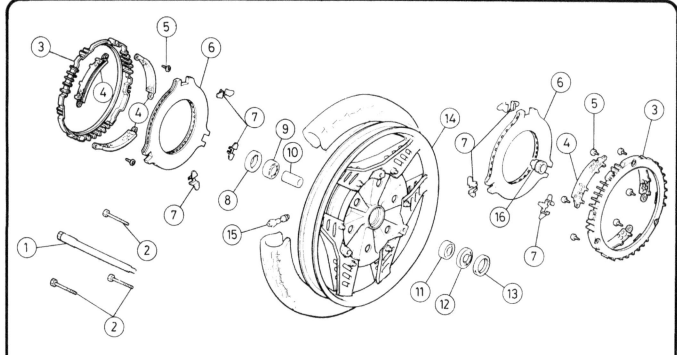

Fig. 5.1 Front wheel

1 Wheel spindle	5 Screw – 12 off	9 Right-hand bearing	13 Oil seal
2 Bolt - 3 off	6 Brake disc - 2 off	10 Spacer	14 Front wheel
3 Brake disc shroud - 2 off	7 Friction spring - 6 off	11 Left-hand bearing	15 Tyre valve
4 Ventilation plate - 6 off	8 Oil seal	12 Speedometer drive plate	16 Speedometer driven gear

3.1a Release screw and free the speedometer drive cable

3.1b Prise off the black plastic caps which cover the screws ...

3.1c ... and remove the caliper anti-dive link

3.1d Remove right-hand caliper locating bolt

3.2a Slacken fork cap nuts and remove the wheel spindle

3.2b Lift the shroud clear and manoeuvre wheel out of forks

4 Front wheel bearings: renewal

1   Remove the front wheel as described in Section 2. Place the wheel on a workbench, supporting it on wooden blocks placed on each side of the bearing boss. Using a screwdriver with the end bent into a slight hook, carefully prise out the oil seals, taking care not to damage the sealing lip if they are to be re-used. Pass a long drift through the upper bearing and lever the spacer to one side. Tap around the inner face of the furthest bearing until it is driven out. Invert the wheel and remove the spacer, then drive out the remaining bearing.
2   Remove all the old grease from the hub and bearings, wash the bearings in petrol, and dry them thoroughly. Check the bearings for roughness by spinning them whilst holding the inner track with one hand and rotating the outer track with the other. If there is the slightest sign of roughness renew them.
3   Before driving the bearings back into the hub, pack the hub with new grease and also grease the bearings. Do not fill the hub more than about $\frac{2}{3}$ full or excess grease will be forced out when the hub heats up in use. Use a tubular drift to drive the bearings back into position. Do not omit to refit the oil seals and distance collar.

5 Rear wheel: removal and replacement

1   Place the machine on its centre stand. Remove the swinging arm end stop bolts, then slacken the wheel spindle nut. Pull the wheel rearwards and swivel the adjusters clear of the end stops, then remove the stops. Push the wheel fully forward and disengage the final drive chain from the rear wheel sprocket.

2   Remove the single bolt which retains the brake hose clamp to the swinging arm. Slacken and remove the three domed nuts which secure the brake shroud to the wheel. Remove the rear wheel spindle and spacer, and rest the wheel on the ground. Manoeuvre the shroud clear of the wheel and along the swinging arm. Pull the wheel rearwards and disengage the complete brake assembly from the wheel. The wheel can now be removed and the brake assembly tied to the swinging arm to support it.
3   The wheel is installed by reversing the removal sequence. It is a good idea to apply a **thin** film of Copaslip or molybdenum disulphide grease to the disc support posts. Before tightening the wheel spindle nut, adjust the final drive chain to give 20 - 25 mm (0.8 - 1.0 in) free play at the centre of the lower run. Check that the adjusters are moved by an equal amount to preserve wheel alignment.

6 Rear wheel bearings: renewal

1   Remove the rear wheel as described in Section 4, and pull off the sprocket and cush drive hub. Other than the absence of an oil seal on the left-hand side of the hub, the bearing arrangement is the same as that described in Section 3, and the same procedure can be adopted. Note that where sealed bearings are encountered, the sealed face should face outwards.
2   To renew the cush drive hub bearing, push out the stepped distance piece, prise out the grease seal and remove the circlip. Invert the hub and drive out the bearing. The new bearing can be driven into place using a large socket. Remember to grease the seal lips prior to installation.

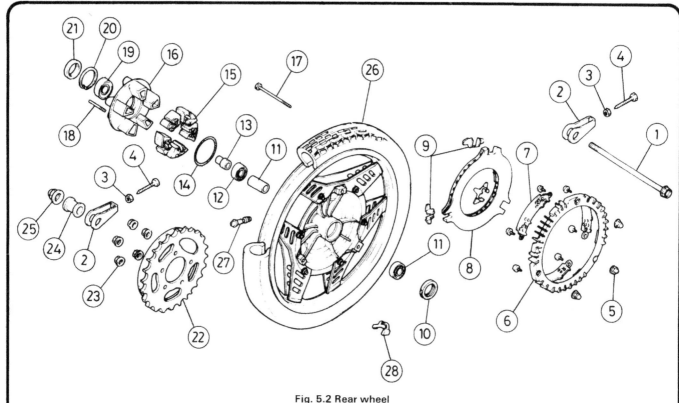

Fig. 5.2 Rear wheel

1 Wheel spindle	8 Brake disc	16 Cush drive hub	22 Final driven sprocket
2 Chain adjuster - 2 off	9 Friction spring - 3 off	17 Bolt - 3 off	23 Nut - 5 off
3 Locknut - 2 off	10 Oil seal	18 Stud - 5 off	24 Left-hand spacer
4 Adjusting bolt - 2 off	11 Spacer	19 Cush drive hub	25 Nut
5 Nut - 3 off	12 Left-hand bearing	bearing	26 Rear wheel
6 Brake disc shroud	13 Shouldered spacer	20 Circlip	27 Tyre valve
7 Ventilation plate	14 O-ring	21 Oil seal	28 Balance weight
- 3 off	15 Cush drive rubbers		

5.2a Release the rear brake hose clamp

5.2b Withdraw the wheel spindle and remove spacer

5.2c Disengage torque stop from swinging arm

5.2d Remove shroud, disengage brake and remove wheel

6.1a Fit right-hand bearing, invert wheel and fit spacer

6.1b Fit bearing with sealed face outwards ..

6.1c ... and drive home using large socket as drift

6.1d Fit grease seal to brake side of hub

6.2a Bearing fits as shown and is retained by circlip

6.2b Grease seal lip and press into housing ...

6.2c ... then install flanged spacer

7   Brake disc and caliper: overhaul

1   Remove the front or rear wheel to free the appropriate brake assembly, referring to Section 2 or 4 for details. Disengage the disc from the caliper and insert a wooden wedge between the brake pads as a precaution against the pads being closed due to accidental operation of the brake.

2   Examine the disc for scoring or damage. The disc will probably appear discoloured due to rusting and the effect of heat on the cast iron. This is normal, but there should be no deep score marks or cracking.

3   If judder under braking has been evident the disc should be checked for distortion using a dial gauge and surface plate. Most owners will find this best entrusted to the dealer or a local engineering shop. Distortion should not exceed 0.3 mm (0.012 in).

4   If the caliper is to be dismantled, connect a length of plastic tubing to the caliper bleed valve and place the other end in a jar. Open the bleed valve and operate the brake lever or pedal until all of the fluid has been expelled, then close the valve and remove the tubing. Remove the special pad retaining screw and lift out the pads and the anti-rattle spring. Pull the brake plate off the caliper and place it to one side. Slacken the hydraulic hose union and remove the caliper from the hose, taking care not to splash any fluid on the paintwork, which will be damaged by it.

5   Clean all brake dust and road dirt from the caliper using

methylated spirit. Place a piece of rag between the caliper halves and displace the pistons by applying an air line or footpump to the hose union. **Do not** place fingers near the pistons; they may be trapped as the pistons are expelled. Pull off the small rubber boots which protect the caliper mounting stud holes. Using an electrical screwdriver, displace and remove the piston seals.

6   The caliper bores and piston outside diameters must be free from any sign of scoring or corrosion. If wear is evident, measure the piston diameters and caliper bores and compare the readings obtained with those shown in the Specifications. **Do not** interchange the pistons.

7   Clean the caliper bores and pistons with new hydraulic fluid, then fit new seals, lubricating these with fluid. Check that the caliper mounting stud holes are clean, then fit new seals, lubricating these with hydraulic fluid. Continue reassembly by reversing the dismantling sequence. Remember to refill and bleed the hydraulic system before using the machine. See Section 9 of this Chapter.

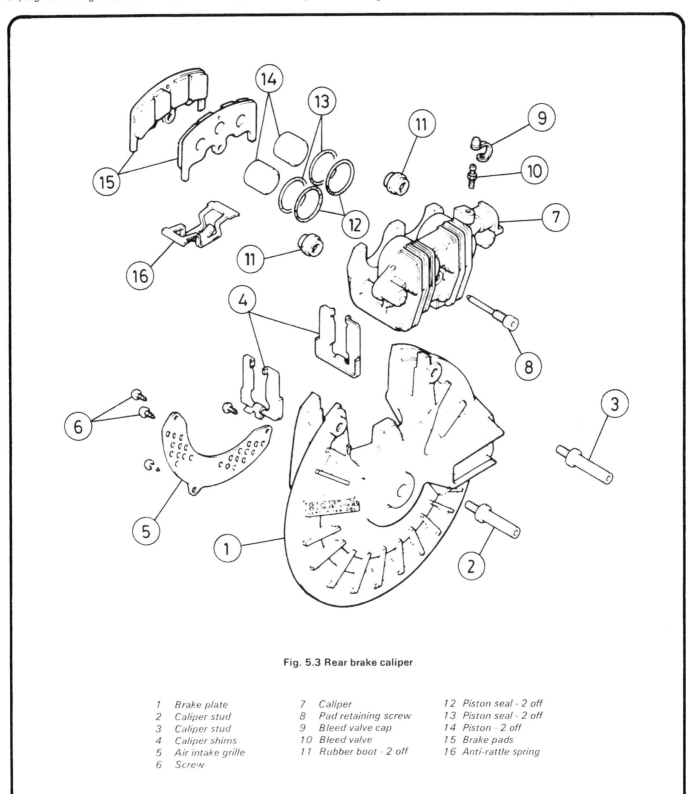

Fig. 5.3 Rear brake caliper

1 Brake plate	7 Caliper	12 Piston seal - 2 off
2 Caliper stud	8 Pad retaining screw	13 Piston seal - 2 off
3 Caliper stud	9 Bleed valve cap	14 Piston - 2 off
4 Caliper shims	10 Bleed valve	15 Brake pads
5 Air intake grille	11 Rubber boot - 2 off	16 Anti-rattle spring
6 Screw		

110

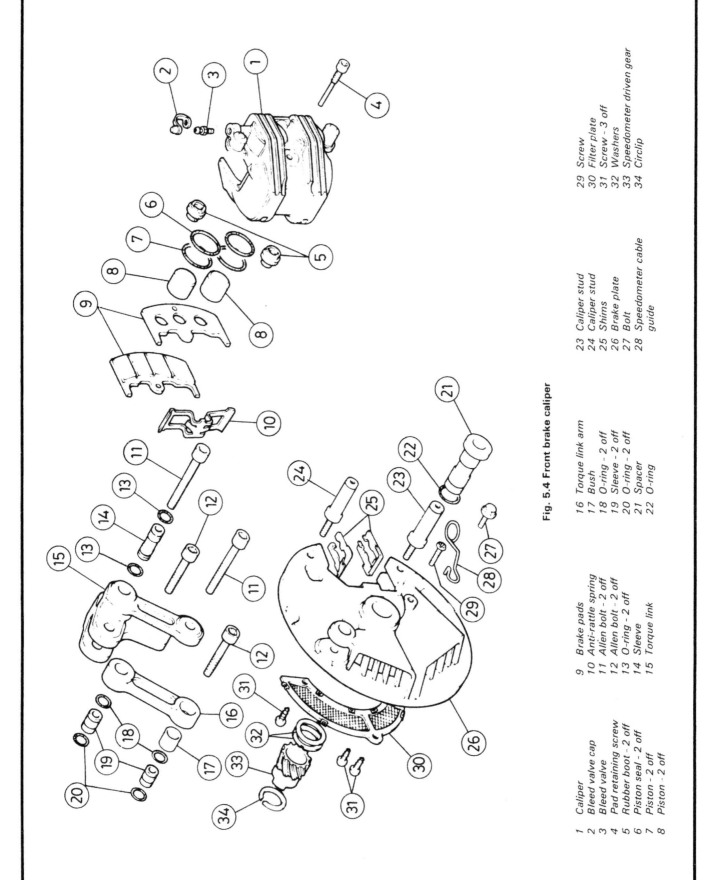

Fig. 5.4 Front brake caliper

1 Caliper
2 Bleed valve cap
3 Bleed valve
4 Pad retaining screw
5 Rubber boot - 2 off
6 Piston seal - 2 off
7 Piston - 2 off
8 Piston - 2 off
9 Brake pads
10 Anti-rattle spring
11 Allen bolt - 2 off
12 Allen bolt - 2 off
13 O-ring - 2 off
14 Sleeve
15 Torque link
16 Torque link arm
17 Bush
18 O-ring - 2 off
19 Sleeve - 2 off
20 O-ring - 2 off
21 Spacer
22 O-ring
23 Caliper stud
24 Caliper stud
25 Shims
26 Brake plate
27 Bolt
28 Speedometer cable guide
29 Screw
30 Filter plate
31 Screw - 3 off
32 Washers
33 Speedometer driven gear
34 Circlip

Tyre changing sequence - tubeless tyres

Deflate tyre. After releasing beads, push tyre bead into well of rim at point opposite valve. Insert lever adjacent to valve and work bead over edge of rim.

Use two levers to work bead over edge of rim. Note use of rim protectors.

When first bead is clear, remove tyre as shown.

Before fitting, ensure that tyre is suitable for wheel. Take note of any sidewall markings such as direction of rotation arrows.

Work first bead over the rim flange.

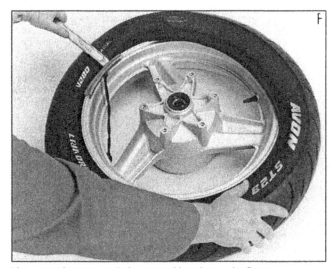

Use a tyre lever to work the second bead over rim flange.

8 Front brake master cylinder: removal and overhaul

1   Drain the hydraulic system as described in Section 6. On machines other than the F2, remove the right-hand handlebar mirror. Remove the single screw which retains the front brake lamp switch to the underside of the lever. Remove the two master cylinder clamp bolts and lift the unit clear of the handlebar. Cover the fuel tank to prevent accidental fluid splashes from damaging the paint, then disconnect the brake hose.

2   Remove the lever pivot bolt locknut and the pivot bolt and place the lever to one side. Pull off the rubber dust seal and release the internal circlip. Displace and remove the secondary cup, piston,

primary cup and spring. If necessary, use an air line or footpump to displace the above.

3   Remove the reservoir cover and clean out any sediment or dust. Check the master cylinder bore surface for scoring or corrosion, and check the piston outside diameter in the same manner. Measure the bore and piston and compare the readings with those shown in the Specifications. If worn, the piston and cylinder must be renewed as an assembly.

4   Lubricate a new set of seals in hydraulic fluid, then assemble the master cylinder by reversing the dismantling sequence. Check that the seal lips enter the bores correctly and are not turned inside out. When refitting the assembled cylinder, note that the top clamp bolt must be tightened first, then the lower bolt. Fill and bleed the hydraulic system as described in Section 9.

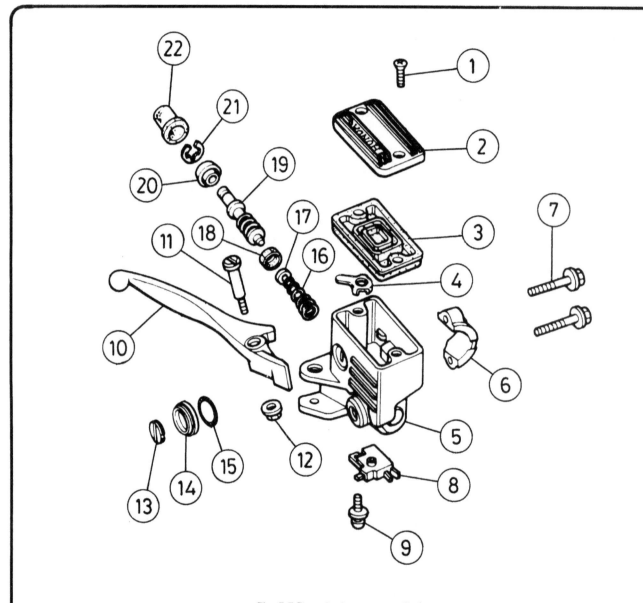

Fig. 5.5 Front brake master cylinder

1 Screw - 2 off	7 Bolt - 2 off	12 Nut	17 Washer
2 Reservoir cap	8 Front brake stop	13 Fluid level sight	18 Primary cup
3 Diaphragm	lamp switch	glass	19 Piston
4 Protection plate	9 Screw and washer	14 Sight glass holder	20 Secondary cup
5 Master cylinder body	10 Brake lever	15 O-ring	21 Circlip
6 Handlebar clamp	11 Lever pivot bolt	16 Spring	22 Dust seal

9 Rear brake master cylinder: removal and overhaul

1   Drain the hydraulic system as described in Section 6. Remove the right-hand side panel. Slacken the hose clip which retains the pipe from the reservoir, then pull off the pipe, catching any residual fluid in some rag. Slacken and remove the union bolt which retains the pipe to the caliper, again catching any residual fluid. Straighten the split pin which secures the clevis pin at the end of the brake arm. The clevis pin can now be displaced to free the operating rod. Remove the two Allen bolts which secure the master cylinder to the footrest plate and lift the master cylinder away.
2   Clean all external surfaces of the master cylinder using methylated spirit. Avoid getting dirt into the reservoir or brake pipe unions. Slide the rubber boot along the pushrod and over the adjuster nut. Do not disturb the adjuster. Release the large internal circlip and remove the pushrod assembly, piston, primary cup and spring. Release the single screw which retains the reservoir to the frame, and clean this and the master cylinder body with new hydraulic fluid.
3   Remove and discard the primary and secondary cups, then check the piston and cylinder bore for scoring or corrosion. Measure the two components, comparing the readings with those given in the Specifications.
4   If the assembly is in serviceable condition, reassemble using a new seal and piston set, lubricating each part with new hydraulic fluid. Refit the reservoir and cylinder, then fill and bleed the system as described in Section 9.

9.1 Rear master cylinder is bolted to inside of the footrest plate

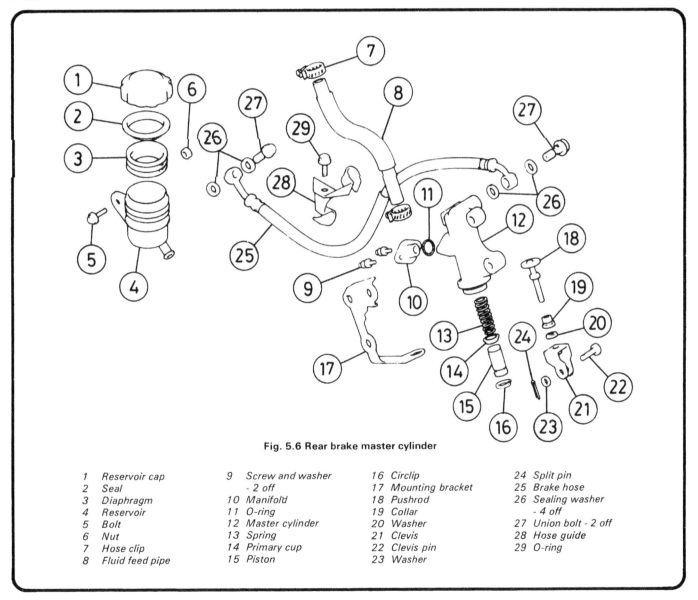

Fig. 5.6 Rear brake master cylinder

1 Reservoir cap	9 Screw and washer	16 Circlip	24 Split pin
2 Seal	- 2 off	17 Mounting bracket	25 Brake hose
3 Diaphragm	10 Manifold	18 Pushrod	26 Sealing washer
4 Reservoir	11 O-ring	19 Collar	- 4 off
5 Bolt	12 Master cylinder	20 Washer	27 Union bolt - 2 off
6 Nut	13 Spring	21 Clevis	28 Hose guide
7 Hose clip	14 Primary cup	22 Clevis pin	29 O-ring
8 Fluid feed pipe	15 Piston	23 Washer	

10 Bleeding the hydraulic brake system

1   If the hydraulic system has to be drained and refilled, if the brake lever or pedal travel becomes excessive or if the brake operates with a soft or spongy feeling, the brakes must be bled to expel air from the system. The procedure for bleeding the hydraulic brakes is best carried out by two persons.
2   First check the fluid level in the reservoir and top up with fresh fluid.
3   Keep the reservoir at least half full of fluid during the bleeding procedure.
4   Fit the cap on to the reservoir to prevent a spout of fluid or the entry of dust into the system. Place a clean glass jar below one caliper bleed screw and attach a clear plastic pipe from the caliper bleed screw to the container. Place some clean hydraulic fluid in the jar so that the pipe is always immersed below the surface of the fluid.
5   Unscrew the bleed screw one half turn and squeeze the brake lever as far as it will go but do not release it until the bleeder valve is closed again. Repeat the operation a few times until no more air bubbles come from the plastic tube.
6   Keep topping up the reservoir with new fluid. When all the bubbles disappear, close the bleeder valve dust cap. Check the fluid level in the reservoir, after the bleeding operation has been completed. Repeat the air bleeding procedure with the second caliper unit (where appropriate).
7   Reinstall the diaphragm and tighten the reservoir cap securely. Do not use the brake fluid drained from the system, since it will contain minute air bubbles.
8   Never use any fluid other than that recommended. Oil must **not be used** under any circumstances.

11 Brake pads: inspection and renewal

1   When carrying out any work on the braking system, it is good practice to check the condition of the pads, in addition to the normal regular maintenance checks. For details refer to the Routine Maintenance Chapter at the front of the manual.

12 Final drive chain, sprockets and cush drive: general

1   These components are responsible for the transmission of power from the engine/gearbox unit to the rear wheel. When in good condition, chain drive is an efficient transmission method, but this efficiency falls off sharply with wear during normal use. It follows that regular maintenance is essential, and reference should be made to Routine Maintenance for details.

13 Wheel balancing

1   It is customary on all high performance machines to balance the wheels complete with tyre and tube. The out of balance forces which exist are eliminated and the handling of the machine is improved in consequence. A wheel which is badly out of balance produces through the steering a most unpleasant hammering effect at high speeds.
2   Some tyres have a balance mark on the sidewall, usually in the form of a coloured spot. This mark must be in line with the tyre valve, when the tyre is fitted to the inner tube. Even then the wheel may require the addition of balance weights, to offset the weight of the tyre valve itself.
3   To overcome drag from the brakes and, in the case of the rear wheel, the final drive components, the wheel should be removed from the machine and set up in an improvised stand to allow it to be spun freely. If the wheel is spun several times, it will probably come to rest with the tyre valve or the heaviest part downward and will always come to rest in the same position. Balance weights must be added at a point diametrically opposite this heavy spot until the wheel will come to rest in ANY position after it is spun.
4   Balance weights can be obtained from Honda dealers in 10 gm, 20 gm and 30 gm sizes, the weights clipping to the rim. Note that the wheels should be rebalanced whenever a new tyre is fitted.

14 Tubeless tyres: removal and replacement

1   It is strongly recommended that should a repair to a tubeless tyre be necessary, the wheel is removed from the machine and taken to a tyre fitting specialist who is willing to do the job, or taken to an official dealer. This is because the force required to break the seal between the wheel rim and tyre bead is considerable and considered to be beyond the capabilities of an individual working with normal tyre removing tools. Any abortive attempt to break the rim to bead seal may also cause damage to the wheel rim, resulting in an expensive wheel replacement. If, however, a suitable bead releasing tool is available, and experience has already been gained in its use, tyre removal and refitting can be accomplished as follows.
2   To remove the tyre from either wheel first detach the wheel from the machine by following the procedure in Sections 3 or 5 depending on whether the front or the rear wheel is involved. Deflate the tyre by removing the valve insert and when it is fully deflated, push the bead of the tyre away from the wheel rim on both sides so that the bead enters the centre well of the rim. As noted, this operation will almost certainly require the use of a bead releasing tool.
3   Insert a tyre lever close to the valve and lever the edge of the tyre over the outside of the wheel rim. Very little force should be necessary; if resistance is encountered it is probably due to the fact that the tyre beads have not entered the well of the wheel rim all the way round the

10.4 Remove dust cap and fit bleed tube to bleed valve

12.1 Front sprocket condition can be checked after removing cover tyre. Should the initial problem persist, lubrication of the tyre bead and the inside edge and lip of the rim will facilitate removal. Use a recommended lubricant, a diluted solution of washing-up liquid or french chalk. Lubrication is usually recommended as an aid to tyre fitting but its use is equally desirable during removal. The risk of lever damage to wheel rims can be minimised by the use of proprietary plastic rim protectors placed over the rim flange at the point where the tyre levers are inserted. Suitable rim protectors may be fabricated very easily from short lengths (4-6 inches) of thick-walled nylon petrol pipe which have been split down one side using a sharp knife. The use of rim protectors should be adopted whenever levers are used and, therefore, when the risk of damage is likely.

4   Once the tyre has been edged over the wheel rim, it is easy to work around the wheel rim so that the tyre is completely free on one side.

5   Working from the other side of the wheel, ease the other edge of the tyre over the outside of the wheel rim, which is furthest away. Continue to work around the rim until the tyre is freed completely from the rim.

6   Refer to the following Section for details relating to puncture repair and the renewal of tyres. See also the remarks relating to the tyre valves in Section 27.

7   Refitting of the tyre is virtually a reversal of removal procedure. If the tyre has a balance mark (usually a spot of coloured paint), as on the tyres fitted as original equipment, this must be positioned alongside the valve. Similarly, any arrow indicating direction of rotation must face the right way.

8   Starting at the point furthest from the valve, push the tyre bead over the edge of the wheel rim until it is located in the central well. Continue to work around the tyre in this fashion until the whole of one side of the tyre is on the rim. It may be necessary to use a tyre lever during the final stages. Here again, the use of a lubricant will aid fitting. It is recommended strongly that when refitting the tyre only a recommended lubricant is used because such lubricants also have sealing properties. Do not be over generous in the application of lubricant or tyre creep may occur.

9   Fitting the upper bead is similar to fitting the lower bead. Start by pushing the bead over the rim and into the well at a point diametrically opposite the tyre valve. Continue working round the tyre, each side of the starting point, ensuring that the bead opposite the working area is always in the well. Apply lubricant as necessary. Avoid using tyre levers unless absolutely essential, to help reduce damage to the soft wheel rim. The use of the levers should be required only when the final portion of bead is to be pushed over the rim.

10 Lubricate the tyre beads again prior to inflating the tyre, and check that the wheel rim is evenly positioned in relation to the tyre beads. Inflation of the tyre may well prove impossible without the use of a high pressure air hose. The tyre will retain air completely only when the beads are firmly against the rim edges at all points and it may be found when using a foot pump that air escapes at the same rate as it is pumped in. This problem may also be encountered when using an air hose on new tyres which have been compressed in storage and by virtue of their profile hold the beads away from the rim edges. To overcome this difficulty, a tourniquet may be placed around the circumference of the tyre, over the central area of the tread. The compression of the tread in this area will cause the beads to be pushed outwards in the desired direction. The type of tourniquet most widely used consists of a length of hose closed at both ends with a suitable clamp fitted to enable both ends to be connected. An ordinary tyre valve is fitted at one end of the tube so that after the hose has been secured around the tyre it may be inflated, giving a constricting effect. Another possible method of seating beads to obtain initial inflation is to press the tyre into the angle between a wall and the floor. With the airline attached to the valve additional pressure is then applied to the tyre by the hand and shin, as shown in the accompanying illustration. The application of pressure at four points around the tyre's circumference whilst simultaneously applying the airhose will often effect an initial seal between the tyre beads and wheel rim, thus allowing inflation to occur.

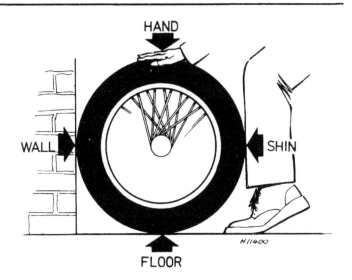

Fig. 5.7 Method of seating tubeless tyre beads

if loss of air is experienced, because there is no inner tube to rupture, in normal conditions a sudden blow-out is avoided.

2   If a puncture of the tyre occurs, the tyre should be removed for inspection for damage before any attempt is made at remedial action. The temporary repair of a punctured tyre by inserting a plug from the outside should not be attempted. Although this type of temporary repair is used widely on cars, the manufacturers strongly recommend that no such repair is carried out on a motorcycle tyre. Not only does the tyre have a thinner carcass, which does not give sufficient support to the plug, but the consequences of a sudden deflation are often sufficiently serious that the risk of such an occurrence should be avoided at all costs.

3   The tyre should be inspected both inside and out for damage to the carcass. Unfortunately the inner lining of the tyre – which takes the place of the inner tube – may easily obscure any damage and some experience is required in making a correct assessment of the tyre condition.

4   There are two main types of tyre repair which are considered safe for adoption in repairing tubeless motorcycle tyres. The first type of repair consists of inserting a mushroom-headed plug into the hole from the inside of the tyre. The hole is prepared for insertion of the plug by reaming and the application of an adhesive. The second repair is carried out by buffing the inner lining in the damaged area and applying a cold or vulcanised patch. Because both inspection and repair, if they are to be carried out safely, require experience in this type of work, it is recommended that the tyre be placed in the hands of a repairer with the necessary skills, rather than repaired in the home workshop.

5   In the event of an emergency, the only recommended 'get-you-home' repair is to fit a standard inner tube of the correct size. If this course of action is adopted, care should be taken to ensure that the cause of the puncture has been removed before the inner tube is fitted. It will be found that the valve hole in the rim is considerably larger than the diameter of the inner tube valve stem. To prevent the ingress of road dirt, and to help support the valve, a spacer should be fitted over the valve. A conversion spacer for most Honda models equipped with Comstar wheels is available from Honda dealers.

6   In the event of the unavailability of tubeless tyres, ordinary tubed tyres fitted with inner tubes of the correct size may be fitted. Refer to the manufacturer or a tyre fitting specialist to ensure that only a tyre and tube of equivalent type and suitability is fitted, and also to advise on the fitting of a valve nut/spacer to the rim hole.

7   **Note:** The Inboard Ventilated Disc Brake fitted to the CBX550 models complicates wheel removal to such an extent that only the most dedicated (or desperate) should consider a roadside repair. Note also that few tyre suppliers welcome tubeless motorcycle tyre repairs, and almost all will insist that the owner removes and refits the wheel. In the event of a puncture away from home, contact the nearest Honda dealer for advice.

15 Puncture repair and tyre renewal

1   The primary advantage of the tubeless tyre is its ability to accept penetration by sharp objects such as nails etc, without loss of air. Even

16 Tyre valves: description and renewal

1   It will be appreciated from the preceding Sections, that the adoption of tubeless tyres has made it necessary to modify the valve arrangement, as there is no longer an inner tube which can carry the valve core. The problem has been overcome by using a moulded rubber valve body which locates in the wheel rim hole. The valve body is pear-shaped, and has a groove around its widest point which engages with the rim forming an airtight seal.

2   The valve is fitted from the rim well, and it follows that it can only be renewed whilst the tyre itself is removed from the wheel. Once the valve has been fitted, it is almost impossible to remove it without damage, and so the simplest method is to cut it as close as possible to the rim well. The two halves of the old valve can then be removed.

3   The new valve is fitted by inserting the threaded end of the valve body through the rim hole, and pulling it through until the groove engages in the rim. In practice, a considerable amount of pressure is required to pull the valve into position, and most tyre fitters have a special tool which screws onto the valve end to enable purchase to be obtained. It is advantageous to apply a little tyre bead lubricant to the valve to ease its insertion. Check that the valve is seated evenly and securely.

4   The incidence of valve body failure is relatively small, and leakage only occurs when the rubber valve case ages and begins to perish. As a precautionary measure, it is advisable to fit a new valve when a new tyre is fitted. This will preclude any risk of the valve failing in service. When purchasing a new valve, it should be noted that a number of different types are available. The correct type for use in the Comstar wheel is a Schrader 413, Bridgeport 183M or equivalent.

5   The valve core is of the same type as that used with tubed tyres, and screws into the valve body. The core can be removed with a small slotted tool which is normally incorporated in plunger type pressure gauges. Some valve dust caps incorporate a projection for removing valve cores. Although tubeless tyre valves seldom give trouble, it is possible for a leak to develop if a small particle of grit lodges on the sealing face. Occasionally, an elusive slow puncture can be traced to a leaking valve core, and this should be checked before a genuine puncture is suspected.

6   The valve dust caps are a significant part of the tyre valve assembly. Not only do they prevent the ingress of road dirt into the valve, but also act as a secondary seal which will reduce the risk of sudden deflation if a valve core should fail.

Chapter 6 Electrical system

Contents

Specifications

Battery
Capacity	12 volt, 12 Ah
Electrolyte specific gravity	1.270 - 1.290 @ 20°C (68°F)
Maximum charge rate	1.2 amperes
Earth	Negative (−)

Alternator
Type	Three-phase, 0.23 kW @ 5000 rpm
Capacity (minimum):	
At 1500 rpm	5A
At 5000 rpm	17A

Regulator/rectifier
Type	Sealed electronic

Starter motor
Type	Four brush
Brush spring tension	800 ± 120 gram (28.2 ± 4.2 oz)
Service limit	680 gram (24.0 oz)
Brush length	12.0 - 13.0 mm (0.47 - 0.51 in)
Service limit	6.5 mm (0.26 in)

Bulbs
Headlamp	12v 60/55W H4 (quartz-halogen)
Tail/brake	12v 5/21W
Turn signal	12v 21W
Parking lamp	12v 4W
Instrument and warning lamps	12v 3.4W

1 General description

The electrical system is supplied by a crankshaft mounted 3-phase alternator. The electromagnetic rotor is secured by a single bolt to the left-hand end of the crankshaft, the stator and brushes being mounted inside the outer cover.

The alternating current (ac) is fed to the system via a combined regulator/rectifier unit which converts it to direct current (dc) and controls the system voltage to a nominal 12 volts.

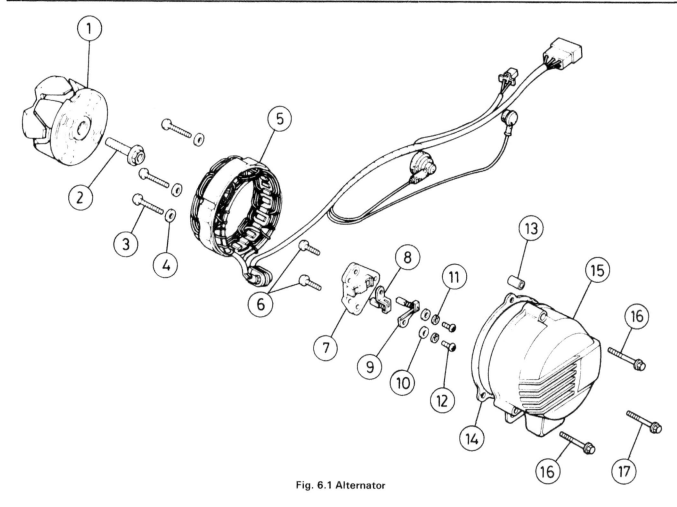

Fig. 6.1 Alternator

1	Rotor	6	Screw - 2 off
2	Bolt	7	Brush plate
3	Screw - 3 off	8	Brush assembly
4	Washer - 3 off	9	Brush assembly
5	Stator	10	Washer - 2 off

11	Spring washer - 2 off	15	Engine left-hand
12	Screw - 2 off		outer cover
13	Locating dowel	16	Bolt - 2 off
14	Cover gasket	17	Bolt

2   Testing the electrical system: general

1   Checking the electrical output and the performance of the various components within the charging system requires the use of test equipment of the multimeter type and also an ammeter of 0-20 ampere range. When carrying out checks, care must be taken to follow the procedures laid down and so prevent inadvertent incorrect connections or short circuits. Irreparable damage to individual components may result if reversal of current or shorting occurs. It is advised that unless some previous experience has been gained in auto-electrical testing the machine be returned to a Honda Service Agent or auto-electrician, who will be qualified to carry out the work and have the necessary test equipment.

2   If the performance of the charging system is suspect, the system as a whole should be checked first, followed by testing of the individual components to isolate the fault. The three main components are the alternator, the rectifier and the regulator. Before commencing the tests, ensure that the battery is fully charged, as described in Routine Maintenance.

3   Charging system: checking the output

1   A quick check of the charging system condition may be made using the above-mentioned ammeter and voltmeter arrangement. Remove the side panels to gain access to the electrical components.

Disconnect the red battery positive (+) lead, attaching it to the positive (+) terminal of the ammeter. Run a lead from the negative (−) ammeter terminal to the positive (+) battery terminal. Set the multimeter to the 0-20 volts dc scale (or higher) and connect the positive (+) probe to the positive (+) battery terminal. The negative (−) probe should be earthed.

2   Start the engine and allow it to reach normal running temperature, then turn on the main lights, set on main beam. At 1500 rpm, the discharge reading found at idle speed should be cancelled out. If the engine speed is now increased to 5000 rpm, the ammeter should show a zero reading or a slight charge, with 14 volts indicated on the voltmeter or multimeter.

3   If the output is erratic or noticeably below the specified amount, either the alternator or the rectifier may be at fault. The rectifier may be tested as described in Section 4. The alternator stator should be tested as follows, using a multimeter set to the resistance function. Disconnect the block connector which connects the three yellow output leads from the alternator to the regulator/rectifier, this being located just below the rear of the fuel tank. Using a multimeter set on the resistance (ohms) scale, check that continuity exists between any pair of yellow leads. Check also that no lead has continuity with earth. If the results of the check do not correspond with those specified, there is evidence of short-circuits or open-circuits in the stator windings or the leads. The specified resistance is not available, but it will be sufficient just to note whether there is an open or short circuit.

4   Low output may also be caused by sticking or worn alternator brushes. These are located in a plastic holder on the inside of the alternator cover. Renew both brushes if either is worn down to or

beyond the scribed wear limit line. The brushes are retained by two screws on the underside of the holder, which can be removed once the stator and brush holder securing screws have been removed and the assembly withdrawn from the cover.

4 Regulator/rectifier unit: location and testing

1   The regulator/rectifier is a finned alloy unit mounted on the frame beneath the fuel tank. In the event of a suspected fault, it can be checked by measuring the resistance between the various leads at the two connector blocks near the rear of the tank. It should be pointed out that the figures quoted in the accompanying table are those that should be indicated when using either a Sanwa SP-10D or Kowa TH-5H tester, and other meters may not produce the same values. In practice, however, any meter should indicate low and infinite resistances, and these will suffice for a general check of the unit.
2   Using the table accompanying this Section, connect the meter test probes as shown and compare the readings obtained. If the reading conflicts on any pair of leads when tested, a diode has failed and it will be necessary to renew the unit. As a precaution, ask your Honda dealer to double-check the diagnosis before ordering a new unit.

4.1 The regulator/recitifer is mounted on the frame top tube

5 Battery: general

1   The battery is of fundamental importance in the electrical system, providing both a reserve of power and a stabilising influence on the various circuits. In the event of any electrical problem, always check the battery condition first, noting that there is little point in checking the system unless the battery is in good condition and fully charged. Refer to Routine Maintenance for details.

6 Starter system: checks

1   In the event of a starter malfunction, always check first that the battery is fully charged. A partly discharged battery may be able to provide enough power for the lighting circuit, but not the very heavy current required for starting the engine.
2   Remove the right-hand side panel and note the location of the starter relay. This is mounted on the rear of the battery tray and can be identified by the two heavy duty cables connected to two of its four terminals. Switch on the ignition and press the starter button. If the relay is operating a distinct click will be heard as the internal solenoid closes the starter lead contact. A silent relay can be assumed to be defunct.
3   Disconnect the heavy duty starter lead at the motor terminal and connect a 12 volt test bulb between it and a sound earth point. Operate the starter switch again. If the bulb lights, the motor is being supplied with power and should be removed for overhaul.

kΩ

+ Probe — Probe	Red/White	Green	Yellow 1	Yellow 2	Yellow 3
Red/White		∞	∞	∞	∞
Green	5~40		5~40	5~40	5~40
Yellow 1	5~40	∞		∞	∞
Yellow 2	5~40	∞	∞		∞
Yellow 3	5~40	∞	∞	∞	

Fig. 6.2 Regulator/rectifier unit – regulator test table

Test results obtained with specified meters – see text

7 Starter motor: removal and overhaul

1   Check that the ignition switch is off and disconnect the battery negative (−) lead to isolate the system. Remove the starter lead at the motor terminal. Release the two motor retaining bolts and remove the motor by pulling it clear of the casing.
2   Remove the two long screws which retain the motor end covers and lift away the brush cover. Ease the brush plate away from the motor casing until the brushes spring clear of the commutator. Disengage the field coil brushes from the brush plate to free it.
3   Remove the end cover and reduction gear assembly then slide the armature out of the motor casing. Note that shims are fitted to both ends of the armature. These must be kept separate and refitted in their original positions.
4   Measure the length of the brushes using a vernier caliper, renewing them as a set if any have worn beyond the 6.5 mm (0.26 in) service limit. If a spring balance or some weights are available, check that the brush springs are able to exert at least 680 gm (24.0 oz) pressure.
5   Clean the commutator segments with methylated spirit and inspect each one for scoring or discolouration. If any pair of segments is discoloured, a shorted armature winding is indicated. The manufacturer supplies no information regarding skimming and re-cutting the armature in the event of serious scoring or burning, and so by implication suggests that a new armature is the only solution. It is suggested, however, that the advice of a vehicle electrical specialist is sought first; professional help may work out a lot cheaper.
6   Honda advise against cleaning the commutator segments with abrasive paper, presumably because of the risk of abrasive particles becoming embedded in the soft segments. It is suggested, therefore, that an ink eraser be used to burnish the segments and remove any surface oxide deposits before installing the brushes.
7   Using a multimeter set on the resistance scale, check the continuity between pairs of segments, noting that anything other than a very low resistance indicates a partially or completely open circuit. Next check the armature insulation by checking for continuity between the armature core and each segment. Anything other than infinite resistance indicates an internal failure.
8   Check the field coil windings by measuring the resistance between each brush and the terminal stud. A high resistance indicates an internal break in the windings. Repeat the test, this time between the brushes and the casing. No reading should be indicated, anything less than infinite resistance suggesting an insulation failure.
9   Reassemble the motor by reversing the dismantling sequence, noting that the reduction gears will benefit from a coating of molybdenum disulphide grease. Make sure that the shims are fitted correctly. Fit the brushes into their holders and eash the brush plate over the commutator, using a screwdriver to work the brushes into place. Note that the motor end covers are located by a slot which engages over a pin on the motor body.

7.2a Remove the starter motor end cover ...

7.2b ... and ease brush plate clear until brushes spring clear

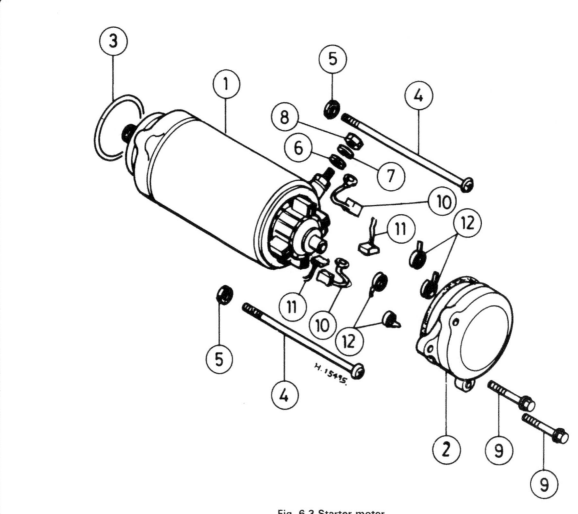

H.15495.

Fig. 6.3 Starter motor

1	Starter motor	5	Spring washer - 2 off	9	Bolt - 2 off
2	Brush cover	6	Washer	10	Brush holder - 2 off
3	O-ring	7	Spring washer	11	Brush - 2 off
4	Screw - 2 off	8	Nut	12	Spring - 4 off

8 Headlamp: bulb renewal and alignment

CB550 F model

1 Remove the two screws which secure the headlamp assembly to the shell, and lift the headlamp clear. Pull off the headlamp bulb connector and pull off the parking lamp assembly.

2 Remove the black dust seal to expose the bulb retainer spring. Disengage the spring, then lift out the bulb. Note that on no account must the bulb's quartz envelope be touched. If this is done accidentally, wipe it with a tissue moistened with methylated spirit.

3 The parking lamp is a push fit in a grommet in the headlamp reflector. The bulb is rated at 12V 4W and has a bayonet fitting. Both bulbs are fitted by reversing the removal sequence.

4 Horizontal alignment should be set so that the beam shines straight ahead on main beam. Adjustment is provided by means of a small screw in the rim edge, and is best set by trial and error during a night-time ride. Vertical alignment is adjusted by slackening the two headlamp mounting bolts and moving the shell up or down as required. In the UK, lighting regulations require that the headlamp is adjusted so that the light will not dazzle a person standing at a distance greater than 25 feet, and whose eye level is not less than 3 ft 6 inches from ground level. Adjustment should be made with the rider seated normally, plus any regular pillion passenger or luggage.

CBX550 F2 model

5 The bulb renewal procedure is as described above, except that the headlamp is removed from the fairing recess by pulling off the adjuster knob, having removed the small screw which secures it, and then releasing the large nut on the adjuster spindle and the adjacent domed nut.

6 Horizontal alignment is adjusted by turning in or out the small screw accessible between the headlamp assembly and fairing recess. Vertical alignment can be corrected whilst riding by turning the large knob inside the fairing.

9 Turn signal and tail/brake lamps: bulb renewal

1 These lamps each employ bayonet fitting bulbs, access to which is gained after removing the lens retaining screws. The tail/brake lamp has a twin filament bulb with offset pins to prevent incorrect fitting.

Fig. 6.4 Headlamp – CBX550 F

1	Outer rim	9	Horizontal alignment
2	Inner rim		screw
3	Reflector	10	Spring
4	Headlamp bulb	11	Washer
5	Retainer spring	12	Nut
6	Dust seal	13	Grommet
7	Screw - 2 off	14	Parking lamp
8	Nut - 2 off	15	Parking lamp
			bulbholder

8.5a Remove screw and remove adjuster knob ...

8.5b ... then remove domed nut and washer

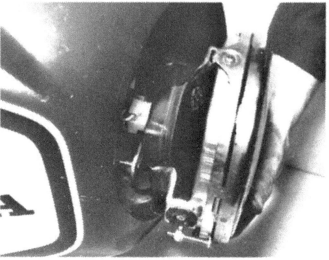

8.5c Lift headlamp assembly clear of fairing recess

8.5d Pull off wiring connector and dust seal

8.5e Disengage bulb retainer spring ends ...

8.5f ... and lift bulb clear of headlamp reflector

8.5g Parking bulb holder is push fit in reflector

9.1a Turn signal bulbs have bayonet fitting

9.1b Tail/brake lamp unit uses twin bulbs

10 Instrument panel: removal and bulb renewal

1   On CBX550 F2 machines, remove the fairing as described in Chapter 4 to allow access to the instrument panel. Unscrew the knurled rings which retain the speedometer and tachometer drive cables. Remove the rubber-bushed screws which retain the panel to the mounting bracket. The instrument panel may now be tilted upwards to allow access to the push-fit bulb holders.

2   If it is necessary to remove the complete panel, make a sketch of the wiring colours to avoid confusion during reassembly. Remove the headlamp and disconnect the instrument panel wiring connector, and push it through the shell cut-out. The panel can be lifted away.

3   In the event that the speedometer, tachometer or fuel gauge heads are to be renewed, separate the instrument panel halves after removing the retaining screws. Reassembly can be carried out by reversing the dismantling sequence.

11 Horn: location and examination

1   The horn is suspended from a flexible steel strip bolted to the frame below the oil cooler.

2   A small screw and locknut arrangement provides adjustment. In the event that the horn's performance deteriorates significantly, experimentation with the screw setting will usually restore it to normal operation.

12 Flash unit: location and replacement

1   The flasher relay unit is located behind the left-hand side panel and is supported on an anti-vibration mounting made of rubber.

2   If the flasher unit is functioning correctly, a series of audible clicks will be heard when the indicator lamps are in action. If the unit malfunctions and all the bulbs are in working order, the usual symptom is one initial flash before the unit goes dead; it will be necessary to replace the unit complete if the fault cannot be attributed to any other cause.

3   Care should be taken when handling the unit to ensure that it is not dropped or otherwise subjected to shocks. The internal components may be irreparably damaged by such treatment.

13 Ignition switch: removal and replacement

1   The combined ignition and lighting master switch is bolted to the upper yoke, and may be removed after the instrument panel has been detached. Disconnect the block connector plug from the base of the ignition switch. The switch is held in place by two screws, after the removal of which the switch can be displaced downwards.

2   Repair of a malfunctioning switch is not practicable as the component is a sealed unit; renewal is the only solution.

3   The switch may be refitted by reversing the dismantling sequence. Remember that a new switch will also require a new set of keys.

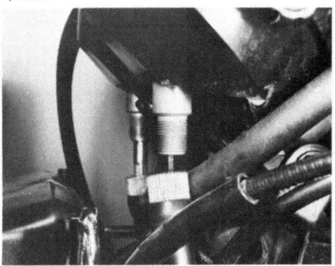

10.1a Disconnect instrument drive cables

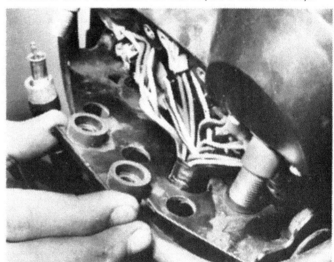

10.1b Release the panel bracket screws to free instruments

10.1c Bulbholders are a push fit in base of panel

12.1 Flasher relay is held by resilient rubber mounting

14.3 Switch contacts can be cleaned with aerosol maintenance fluid

14 Handlebar switches: general

1   Generally speaking, the switches give little trouble but if necessary they can be dismantled by separating the halves which form a split clamp around the handlebars. Note that the machine cannot be started until the ignition cut-out on the right-hand end of the handlebars is turned to the central 'ON' position.
2   Always disconnect the battery before removing any of the switches, to prevent the possibility of a short circuit. Most troubles are caused by dirty contacts but in the event of the breakage of some internal part, it will be necessary to renew the complete switch.
3   Because the internal components of each switch are very small, and therefore difficult to dismantle and reassemble, it is suggested a special electrical contact cleaner be used to clean corroded contacts. This can be sprayed into each switch, without the need for dismantling.

15 Neutral switch: general

1   A small switch fitted to the left-hand side of the crankcase operates a warning lamp in the instrument panel to indicate that neutral has been selected. More importantly, it is interconnected with the starter solenoid and will only allow the engine to be started if the gearbox is in neutral, unless the clutch is disengaged. It can be checked by setting a multimeter on the resistance scale and connecting one probe to the switch terminal and the other to earth. The meter should indicate continuity when neutral is selected and infinite resistance when in any gear.

16 Switch testing procedure

1   In the event of a suspected malfunction the various switch contacts can be checked for continuity in a similar manner to that described in the preceding Section. Details of the switch connections and the appropriate wiring will be found in the wiring diagram that follow. Note that the electrical system should be isolated by disconnecting the battery leads to avoid short circuits.

17 Stop lamp switch: adjustment

1   All models have a stop lamp switch fitted to operate in conjunction with the rear brake pedal. The switch is located immediately above the

swinging arm on the right-hand side of the machine, it has a threaded body giving a range of adjustment.
2   If the stop lamp is late in operating, hold the body still and rotate the combined adjusting/mounting nut in an anti-clockwise direction so that the body moves away from the brake pedal shaft. If the switch operates too early or has a tendency to stick on, rotate the nut in a clockwise direction. As a guide, the light should operate after the brake pedal has been depressed by about 2 cm ($\frac{3}{4}$ inch).
3   The front brake lever incorporates a small switch retained by a single screw to the underside of the switch housing. The switch is a sealed unit and must be renewed if it is found to be defective.

18 Clutch interlock switch: general

1   A small plunger-type switch is incorporated in the clutch lever, serving to prevent operation of the starter circuit when any gear has been selected, unless the clutch lever is held in. It can be checked by the method described above for the neutral switch. If defective it must be renewed, as there is no satisfactory means of repair. The switch can be removed after releasing the clutch cable and lever blade.

19 Fuel level gauge and sender: testing

1   The fuel level sender unit consists of a variable resistor controlled by a float inside the fuel tank. The unit can be unbolted and removed from the tank after the latter has been removed, drained and inverted. See Chapter 2 for details.
2   Connect a multimeter to the sender leads and set it to the resistance scale, then note the readings with the float in the full and empty positions. These should be 4 – 10 ohms and 90 – 100 ohms respectively.
3   To check the gauge itself, first check the sender as described above, then connect it to its wiring connector and switch on the ignition. If the gauge does not move in response to movement of the float arm, the gauge can be considered faulty.

20 Oil pressure switch: testing

1   The oil pressure switch should show continuity when oil pressure falls below 2.8 psi, and no continuity above 2.8 – 5.6 psi. If the oil pressure light comes on whilst riding, **stop immediately** in case of pressure loss. It is best to check a suspect switch by substituting a new one. The switch screws into the external oil take-off near the starter motor. When fitting a new switch use new sealing washers and a sealing compound on the switch threads.

21 Fuses: location

1   The main fuse is a ribbon type element rated at 30 amps. It is housed, together with a spare, in a small black holder next to the rear brake reservoir.
2   Further fuses protect the various circuits, these being housed below a cover on the upper yoke for easy access. Fuses are fitted to protect the electrical system in the event of a short circuit or sudden surge; they are, in effect, an intentional 'weak link', which will blow, in preference to the circuit burning out.
3   Before replacing a fuse that has blown, check that no obvious short circuit has occurred, otherwise the replacement fuse will blow immediately it is inserted. It is always wise to check the electrical circuit thoroughly, to trace the fault and eliminate it.
4   When a fuse blows while the machine is running and no spare is available, a 'get you home' remedy is to remove the blown fuse and wrap it in silver paper before replacing it in the fuseholder. The silver paper will restore the electrical continuity by bridging the broken fuse wire. This expedient should **never** be used if there is evidence of a short circuit or other major electrical fault, otherwise more serious damage will be caused. Replace the 'doctored' fuse at the earliest possible opportunity, to restore full circuit protection. It follows that spare fuses that are used should be replaced as soon as possible to prevent the above situation from arising.

17.2 Rear brake switch can be adjusted by turning nut

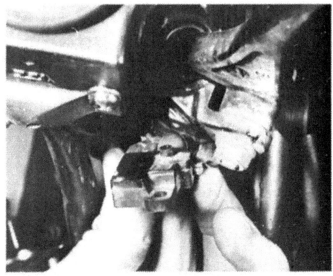

17.3 Front brake switch is screwed to underside of lever

19.1a Release the four nuts which hold the sender ...

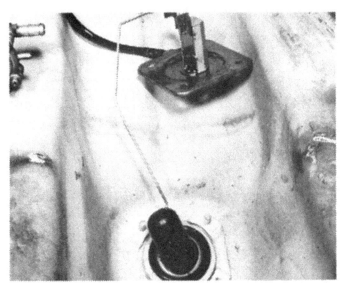

19.1b ... and lift unit clear of tank

20.1 The oil pressure switch is located to the rear of the starter motor

21.1 Main fuse is a 30 Amp ribbon type   21.2 Circuit fuses are housed under handlebar console

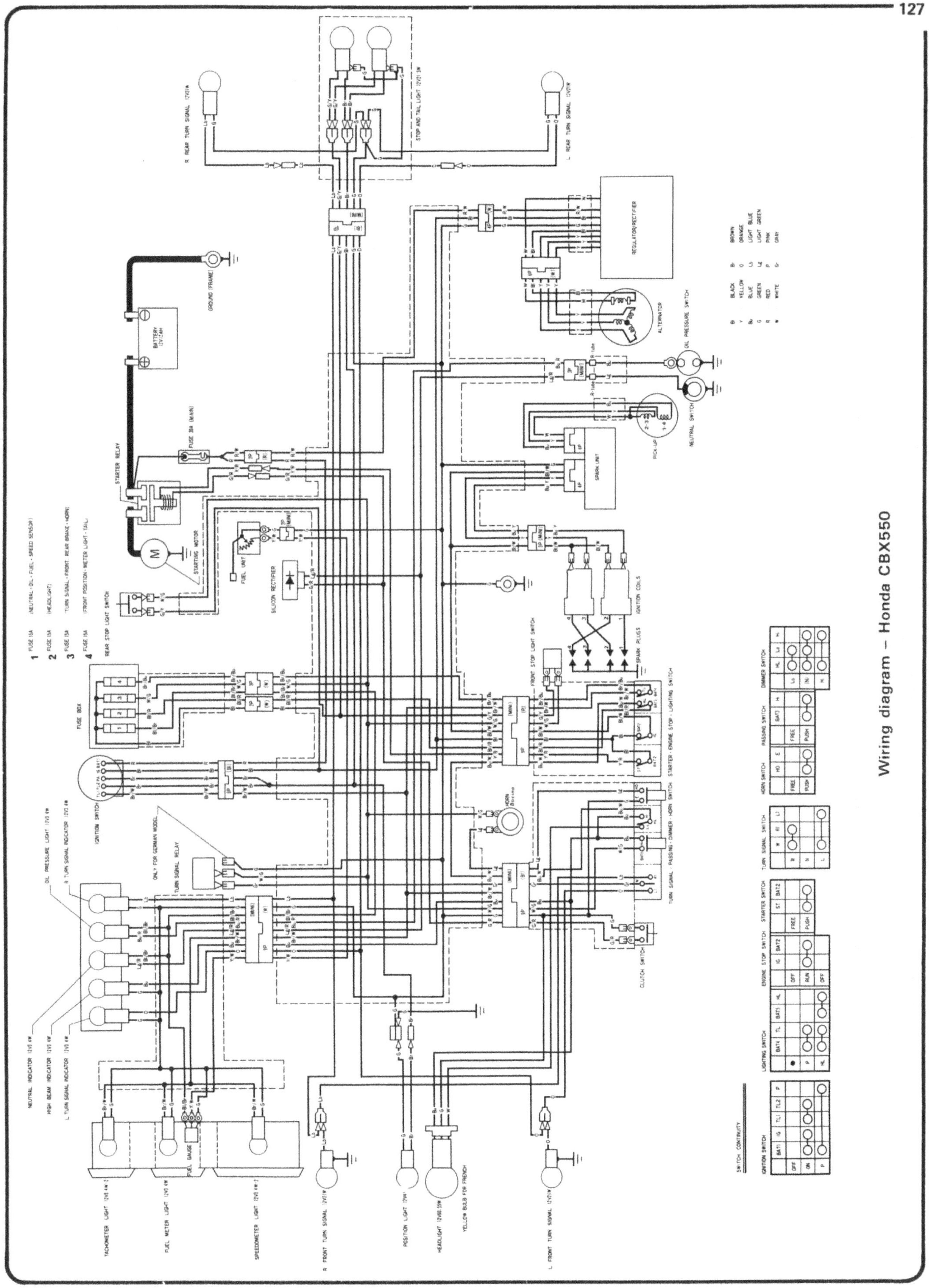

Wiring diagram – Honda CBX550

REAR TURN SIGNAL (12V10W)

STOP AND TAIL LIGHT (12V21/5W)

L. REAR TURN SIGNAL (12V10W)

REGULATOR/RECTIFIER

ALTERNATOR

OIL PRESSURE SWITCH

NEUTRAL SWITCH

SPARK UNIT

PICK UP

GROUND (FRAME)

BATTERY (12V12AH)

FUSE 30A (MAIN)

STARTER RELAY

STARTING MOTOR

FUEL UNIT

SILICON RECTIFIER

REAR STOP LIGHT SWITCH

IGNITION COILS

SPARK PLUGS

FRONT STOP LIGHT SWITCH

FUSE BOX

1 FUSE 15A (NEUTRAL - OIL - FUEL - SPEED SENSOR)
2 FUSE 15A (HEADLIGHT)
3 FUSE 15A (TURN SIGNAL - FRONT - REAR BRAKE - HORN)
4 FUSE 15A (FRONT POSITION METER LIGHT - TAIL)

IGNITION SWITCH

OIL PRESSURE LIGHT

R. TURN SIGNAL INDICATOR (12V3.4W)

ONLY FOR GERMAN MODEL

TURN SIGNAL RELAY

HORN

CLUTCH SWITCH

STARTER STOP - LIGHTING SWITCH

TURN SIGNAL PASSING DIMMER HORN SWITCH

NEUTRAL INDICATOR

HIGH BEAM INDICATOR (12V3.4W)

L. TURN SIGNAL INDICATOR (12V3.4W)

TACHOMETER LIGHT (12V3.4W)

FUEL METER LIGHT (12V3.4W)

FUEL GAUGE

SPEEDOMETER LIGHT (12V3.4W)

R. FRONT TURN SIGNAL (12V10W)

POSITION LIGHT (12V4W)

HEADLIGHT (12V60/55W)

YELLOW BULB FOR FRENCH

L. FRONT TURN SIGNAL (12V10W)

BROWN   Br
ORANGE   O
LIGHT BLUE   Lb
LIGHT GREEN Lg
PINK   P
GRAY   Gr

BLACK   Bl
YELLOW   Y
BLUE   Bu
GREEN   G
RED   R
WHITE   W

SWITCH CONTINUITY

DIMMER SWITCH

	HL	Lo	Hi

PASSING SWITCH

	BAT5	
FREE		
PUSH		

TURN SIGNAL SWITCH

	W	R	L
R			
N			
L			

HORN SWITCH

	HO	E
FREE		
PUSH		

STARTER SWITCH

	ST	BAT2
FREE		
PUSH		

ENGINE STOP SWITCH

	IG	BAT2
OFF		
RUN		
OFF		

LIGHTING SWITCH

	BAT1	TL	HL
OFF			
P			
HL			

IGNITION SWITCH

	BAT1	IG	TL	TL2	P
OFF					
ON					
P					

Metric conversion tables

Inches	Decimals	Millimetres		Millimetres to Inches		Inches to Millimetres
			mm	Inches	Inches	mm
1/64	0.015625	0.3969	0.01	0.00039	0.001	0.0254
1/32	0.03125	0.7937	0.02	0.00079	0.002	0.0508
3/64	0.046875	1.1906	0.03	0.00118	0.003	0.0762
1/16	0.0625	1.5875	0.04	0.00157	0.004	0.1016
5/64	0.078125	1.9844	0.05	0.00197	0.005	0.1270
3/32	0.09375	2.3812	0.06	0.00236	0.006	0.1524
7/64	0.109375	2.7781	0.07	0.00276	0.007	0.1778
1/8	0.125	3.1750	0.08	0.00315	0.008	0.2032
9/64	0.140625	3.5719	0.09	0.00354	0.009	0.2286
5/32	0.15625	3.9687	0.1	0.00394	0.01	0.254
11/64	0.171875	4.3656	0.2	0.00787	0.02	0.508
3/16	0.1875	4.7625	0.3	0.01181	0.03	0.762
13/64	0.203125	5.1594	0.4	0.01575	0.04	1.016
7/32	0.21875	5.5562	0.5	0.01969	0.05	1.270
15/64	0.234375	5.9531	0.6	0.02362	0.06	1.524
1/4	0.25	6.3500	0.7	0.02756	0.07	1.778
17/64	0.265625	6.7469	0.8	0.03150	0.08	2.032
9/32	0.28125	7.1437	0.9	0.03543	0.09	2.286
19/64	0.296875	7.5406	1	0.03937	0.1	2.54
5/16	0.3125	7.9375	2	0.07874	0.2	5.08
21/64	0.328125	8.3344	3	0.11811	0.3	7.62
11/32	0.34375	8.7312	4	0.15748	0.4	10.16
23/64	0.359375	9.1281	5	0.19685	0.5	12.70
3/8	0.375	9.5250	6	0.23622	0.6	15.24
25/64	0.390625	9.9219	7	0.27559	0.7	17.78
13/32	0.40625	10.3187	8	0.31496	0.8	20.32
27/64	0.421875	10.7156	9	0.35433	0.9	22.86
7/16	0.4375	11.1125	10	0.39370	1	25.4
29/64	0.453125	11.5094	11	0.43307	2	50.8
15/32	0.46875	11.9062	12	0.47244	3	76.2
31/64	0.484375	12.3031	13	0.51181	4	101.6
1/2	0.5	12.7000	14	0.55118	5	127.0
33/64	0.515625	13.0969	15	0.59055	6	152.4
17/32	0.53125	13.4937	16	0.62992	7	177.8
35/64	0.546875	13.8906	17	0.66929	8	203.2
9/16	0.5625	14.2875	18	0.70866	9	228.6
37/64	0.578125	14.6844	19	0.74803	10	254.0
19/32	0.59375	15.0812	20	0.78740	11	279.4
39/64	0.609375	15.4781	21	0.82677	12	304.8
5/8	0.625	15.8750	22	0.86614	13	330.2
41/64	0.640625	16.2719	23	0.09551	14	355.6
21/32	0.65625	16.6687	24	0.94488	15	381.0
43/64	0.671875	17.0656	25	0.98425	16	406.4
11/16	0.6875	17.4625	26	1.02362	17	431.8
45/64	0.703125	17.8594	27	1.06299	18	457.2
23/32	0.71875	18.2562	28	1.10236	19	482.6
47/64	0.734375	18.6531	29	1.14173	20	508.0
3/4	0.75	19.0500	30	1.18110	21	533.4
49/64	0.765625	19.4469	31	1.22047	22	558.8
25/32	0.78125	19.8437	32	1.25984	23	584.2
51/64	0.796875	20.2406	33	1.29921	24	609.6
13/16	0.8125	20.6375	34	1.33858	25	635.0
53/64	0.828125	21.0344	35	1.37795	26	660.4
27/32	0.84375	21.4312	36	1.41732	27	685.8
55/64	0.859375	21.8281	37	1.4567	28	711.2
7/8	0.875	22.2250	38	1.4961	29	736.6
57/64	0.890625	22.6219	39	1.5354	30	762.0
29/32	0.90625	23.0187	40	1.5748	31	787.4
59/64	0.921875	23.4156	41	1.6142	32	812.8
15/16	0.9375	23.8125	42	1.6535	33	838.2
61/64	0.953125	24.2094	43	1.6929	34	863.6
31/32	0.96875	24.6062	44	1.7323	35	889.0
63/64	0.984375	25.0031	45	1.7717	36	914.4

Conversion factors

Length (distance)

Inches (in)	X	25.4	= Millimetres (mm)	X 0.0394	= Inches (in)
Feet (ft)	X	0.305	= Metres (m)	X 3.281	= Feet (ft)
Miles	X	1.609	= Kilometres (km)	X 0.621	= Miles

Volume (capacity)

Cubic inches (cu in; in^3)	X	16.387	= Cubic centimetres (cc; cm^3)	X 0.061	= Cubic inches (cu in; in^3)
Imperial pints (Imp pt)	X	0.568	= Litres (l)	X 1.76	= Imperial pints (Imp pt)
Imperial quarts (Imp qt)	X	1.137	= Litres (l)	X 0.88	= Imperial quarts (Imp qt)
Imperial quarts (Imp qt)	X	1.201	= US quarts (US qt)	X 0.833	= Imperial quarts (Imp qt)
US quarts (US qt)	X	0.946	= Litres (l)	X 1.057	= US quarts (US qt)
Imperial gallons (Imp gal)	X	4.546	= Litres (l)	X 0.22	= Imperial gallons (Imp gal)
Imperial gallons (Imp gal)	X	1.201	= US gallons (US gal)	X 0.833	= Imperial gallons (Imp gal)
US gallons (US gal)	X	3.785	= Litres (l)	X 0.264	= US gallons (US gal)

Mass (weight)

Ounces (oz)	X	28.35	= Grams (g)	X 0.035	= Ounces (oz)
Pounds (lb)	X	0.454	= Kilograms (kg)	X 2.205	= Pounds (lb)

Force

Ounces-force (ozf; oz)	X	0.278	= Newtons (N)	X 3.6	= Ounces-force (ozf; oz)
Pounds-force (lbf; lb)	X	4.448	= Newtons (N)	X 0.225	= Pounds-force (lbf; lb)
Newtons (N)	X	0.1	= Kilograms-force (kgf; kg)	X 9.81	= Newtons (N)

Pressure

Pounds-force per square inch (psi; lbf/in^2; lb/in^2)	X	0.070	= Kilograms-force per square centimetre (kgf/cm^2; kg/cm^2)	X 14.223	= Pounds-force per square inch (psi; lbf/in^2; lb/in^2)
Pounds-force per square inch (psi; lbf/in^2; lb/in^2)	X	0.068	= Atmospheres (atm)	X 14.696	= Pounds-force per square inch (psi; lbf/in^2; lb/in^2)
Pounds-force per square inch (psi; lbf/in^2; lb/in^2)	X	0.069	= Bars	X 14.5	= Pounds-force per square inch (psi; lbf/in^2; lb/in^2)
Pounds-force per square inch (psi; lbf/in^2; lb/in^2)	X	6.895	= Kilopascals (kPa)	X 0.145	= Pounds-force per square inch (psi; lbf/in^2; lb/in^2)
Kilopascals (kPa)	X	0.01	= Kilograms-force per square centimetre (kgf/cm^2; kg/cm^2)	X 98.1	= Kilopascals (kPa)

Torque (moment of force)

Pounds-force inches (lbf in; lb in)	X	1.152	= Kilograms-force centimetre (kgf cm; kg cm)	X 0.868	= Pounds-force inches (lbf in; lb in)
Pounds-force inches (lbf in; lb in)	X	0.113	= Newton metres (Nm)	X 8.85	= Pounds-force inches (lbf in; lb in)
Pounds-force inches (lbf in; lb in)	X	0.083	= Pounds-force feet (lbf ft; lb ft)	X 12	= Pounds-force inches (lbf in; lb in)
Pounds-force feet (lbf ft; lb ft)	X	0.138	= Kilograms-force metres (kgf m; kg m)	X 7.233	= Pounds-force feet (lbf ft; lb ft)
Pounds-force feet (lbf ft; lb ft)	X	1.356	= Newton metres (Nm)	X 0.738	= Pounds-force feet (lbf ft; lb ft)
Newton metres (Nm)	X	0.102	= Kilograms-force metres (kgf m; kg m)	X 9.804	= Newton metres (Nm)

Power

Horsepower (hp)	X	745.7	= Watts (W)	X 0.0013	= Horsepower (hp)

Velocity (speed)

Miles per hour (miles/hr; mph)	X	1.609	= Kilometres per hour (km/hr; kph)	X 0.621	= Miles per hour (miles/hr; mph)

Fuel consumption*

Miles per gallon, Imperial (mpg)	X	0.354	= Kilometres per litre (km/l)	X 2.825	= Miles per gallon, Imperial (mpg)
Miles per gallon, US (mpg)	X	0.425	= Kilometres per litre (km/l)	X 2.352	= Miles per gallon, US (mpg)

Temperature

Degrees Fahrenheit = (°C x 1.8) + 32

Degrees Celsius (Degrees Centigrade; °C) = (°F - 32) x 0.56

*It is common practice to convert from miles per gallon (mpg) to litres/100 kilometres (l/100km), where mpg (Imperial) x l/100 km = 282 and mpg (US) x l/100 km = 235

Index